SOVIET MAN IN SPACE

University Press of the Pacific
Honolulu, Hawaii

Soviet Man in Space

by
Yuri Gagarin, et al

ISBN: 0-89875-460-7

Copyright © 2001 by University Press of the Pacific

Reprinted from the original edition

University Press of the Pacific
Honolulu, Hawaii
http://www.universitypressofthepacific.com

All rights reserved, including the right to reproduce this book, or portions thereof, in any form.

In order to make original editions of historical works available to scholars at an economical price, this facsimile of the original edition is reproduced from the best available copy and has been digitally enhanced to improve legibility, but the text remains unaltered to retain historical authenticity.

CONTENTS

Great Event 7

   Highlights of the Hero's Life Story 7

Man's First Space Flight. *A TASS Report* 9

   Man's Safe Return from His First Space Flight ... 10

The Soviet Union Ushers In a New Era in Human Progress 12

Message from the Central Committee of the C.P.S.U., the Presidium of the Supreme Soviet of the U.S.S.R. and the Soviet Government to the Communist Party and the Peoples of the Soviet Union, to the Peoples and Governments of All Countries, to the Whole of Progressive Mankind 12

Statement Made by Y. A. Gagarin Before the Take-Off 15

Spaceport 17

Report from the Landing Area of the **Vostok** 20

To Major **Yuri Alexeyevich Gagarin**, the Soviet Cosmonaut Who Was the First in the World to Accomplish a Space Flight 23

"A Feat that Will Live Through the Ages." *Conversation Between N. S. Khrushchov and Y. A. Gagarin, the First Space Pilot* 24

Space Pilot Speaking 28

"This Feat Is an Embodiment of the Genius of the Soviet People and the Great Might of Socialism" 35

Nation Hails the Hero

    Pioneer of the Universe Greeted 38

    Meeting and Demonstration in Red Square 41

    Speech by Y. A. Gagarin 42

        The Great Feat Will Go Down in the Ages. *Speech by N. S. Khrushchov* 45

Decree of the Presidium of the Supreme Soviet of the U.S.S.R. Conferring the Title of Hero of the Soviet Union upon the Soviet Pilot-Cosmonaut Major Y. A. Gagarin, the World's First Space Flyer 55

Decree of the Presidium of the Supreme Soviet of the U.S.S.R. Instituting the Title of Pilot-Cosmonaut of the U.S.S.R. 56

Decree of the Presidium of the Supreme Soviet of the U.S.S.R. Conferring the Title of Pilot-Cosmonaut of the U.S.S.R. upon Major of the Air Force Y. A. Gagarin 56

Triumph of Labour, Science and Reason. Press Conference at the House of Scientists 57

    Speech by A. N. Nesmeyanov 57

    Speech by Y. A. Gagarin 61

    Question Period 66

Man's First Flight into Space 71
  Decisive Step in Mastering Space 72
  The Spaceship **Vostok** 75
  Medical and Biological Problems of Man's Flight into
    Space. 83
  Training Space Pilots 86
  First Space Flight 91

GREAT EVENT

The first manned space flight in history was accomplished on April 12, 1961, when the Soviet spaceship *Vostok* (East) orbited the earth and made a safe landing.

The first man in space was Yuri Alexeyevich Gagarin, a citizen of the Union of Soviet Socialist Republics.

HIGHLIGHTS OF THE HERO'S LIFE STORY

Major Yuri Gagarin, the first spaceman, is 27.

He was born in Gzhatsk District, Smolensk Region (in the Russian Federation), on March 9, 1934. His father is a collective farmer.

He entered school in 1941, but his schooling was interrupted by the Nazi invasion.

After the war the Gagarin family moved to the town of Gzhatsk, where Yuri continued his studies in secondary school. In 1951 he finished a trade school in the town of Lyubertsy, near Moscow, with honours, qualifying as a foundryman. He finished an evening school for young workers at the same time.

From there he went on to an industrial school in Saratov on the Volga, which he finished with honours in 1955.

While attending the industrial school he took up flying in his spare time at the Saratov flying club. After finishing the course at the club in 1955, he studied at a flying school in Orenburg. He has been a flyer since 1957, when he graduated with honours from the Orenburg school.

Yuri Gagarin became a member of the Communist Party of the Soviet Union in 1960.

His wife Valentina, 26, is a graduate of the Orenburg medical school. They have two girls, Yelena, aged two, and Galya, only one month old.

Yuri's father, now 59, is a carpenter. His mother Anna, 58, is a housewife.

MAN'S FIRST SPACE FLIGHT

A TASS REPORT

On April 12, 1961, in the Soviet Union, the world's first satellite spaceship *Vostok*, with a man on board, was put into orbit round the earth.

The pilot of the *Vostok* is Major of the Air Force Yuri Alexeyevich GAGARIN, a citizen of the Union of Soviet Socialist Republics.

After successful launching in the multi-stage space rocket the satellite ship, having attained orbital velocity and separated from the last stage of the carrier-rocket, had begun free orbital flight round the earth.

According to preliminary data, orbital period of the spaceship is 89.1 minutes; its minimum distance from the earth's surface (perigee) is 175 kilometres and its maximum (apogee), 302 kilometres; the orbit is inclined to the equator at 65°4'.

Together with its pilot, the spaceship weighs 4,725 kilograms excluding the weight of the last stage of the launching rocket.

Two-way radio communication has been established, and is being maintained, with the spaceman, Comrade GAGARIN. The ship's short-wave transmitters are operating on 9.019 megacycles and 20.006 megacycles, and on 143.625 megacycles in the ultrashort-wave band. The condition of the space pilot during flight is being observed by means of radiotelemetering and television systems.

Comrade GAGARIN, the space pilot, withstood the period of acceleration satisfactorily and at present feels quite well. The systems ensuring the necessary life conditions in the cabin of the spaceship are functioning normally.

The flight of the *Vostok* with Comrade GAGARIN on board continues.

* * *

0952 HOURS

According to information received from the spaceship *Vostok*, the pilot, Major Cagarin, reported at 0952 hours Moscow Time as he flew over South America: "Flight proceeding normally, am feeling fine."

1015 HOURS

At 1015 hours Moscow Time Major Gagarin, the space pilot, reported from the spaceship *Vostok*, while flying over Africa: "Flight proceeding normally, am feeling no ill effects from weightlessness."

1025 HOURS

At 1025 hours Moscow Time, after the flight round the globe had been carried out in accordance with the pre-set programme, the deceleration system was switched on and the spaceship with Major Gagarin, the space pilot, on board began to descend from orbit to land in a predetermined area of the Soviet Union.

MAN'S SAFE RETURN FROM HIS FIRST SPACE FLIGHT

After carrying out the planned investigations and the assigned flight programme, the Soviet spaceship *Vostok* made a safe landing in a predetermined area of the Soviet Union on April 12, 1961, at 1055 hours Moscow Time.

The space pilot, Major Gagarin, reported: "PLEASE REPORT TO PARTY AND GOVERNMENT AND TO NIKITA SERGEYEVICH KHRUSHCHOV IN PERSON THAT LANDING WENT OFF NORMALLY, I AM ALL RIGHT AND HAVE NO INJURIES OR BRUISES."

The accomplishment of a manned space flight holds out vast prospects for man's conquest of space.

The Soviet Union Ushers In a New Era in Human Progress

MESSAGE

FROM THE CENTRAL COMMITTEE OF THE C.P.S.U., THE PRESIDIUM OF THE SUPREME SOVIET OF THE U.S.S.R. AND THE SOVIET GOVERNMENT

**TO THE COMMUNIST PARTY
AND THE PEOPLES OF THE SOVIET UNION
TO THE PEOPLES AND GOVERNMENTS OF ALL COUNTRIES
TO THE WHOLE OF PROGRESSIVE MANKIND**

A great event has taken place: for the first time in history, man has accomplished a space flight.

On April 12, 1961, at 0907 hours Moscow Time, the satellite spaceship *Vostok*, with a man on board, went up into space, and after circling the globe, safely returned to the sacred soil of our country, the Land of Soviets.

The first man to have penetrated into space is a Soviet man, a citizen of the Union of Soviet Socialist Republics.

It is an unparalleled victory of man over the forces of Nature, an immense achievement of science and technology, and a triumph for the human mind. It has led off man's flights into space.

This feat, which will live through the ages, is an embodiment of the genius of the Soviet people and the great might of socialism.

The Central Committee of the Communist Party, the Presidium of the Supreme Soviet of the U.S.S.R. and the Soviet Government note with deep satisfaction and legitimate pride that this new era in the progressive development of mankind has been ushered in by our country, the country of victorious socialism.

Tsarist Russia was a backward country and could never have dreamed of accomplishing such feats in the struggle for progress or of competing with technically and economically more developed countries.

The working class and the people as a whole, inspired by the Communist Party led by Lenin, willed that our country should become a mighty socialist power and should attain unprecedented heights in science and technology.

When, in October 1917, the working class took power into its own hands, there were many people, even among the fair-minded, who doubted whether it would be able to govern the country and at least maintain the economic, scientific and technological standards already achieved.

Today the Soviet Union's working class, collective-farm peasantry and intelligentsia, the Soviet people as a whole, are demonstrating an unprecedented victory of science and technology. Our country has surpassed all the other countries of the world by blazing the first trail into space.

The Soviet Union was the first to launch an intercontinental ballistic missile, the first to put a man-made earth satellite into orbit, the first to send a spaceship to the moon. It made the first artificial satellite of the sun, and launched a spaceship towards the planet Venus. One after another, Soviet spaceships with living creatures on board went into orbit and returned to earth.

The triumphant flight of a Soviet man round the earth in a spaceship was a victory crowning our exploration of space.

Honour and glory to the working class, the Soviet peasantry, and the Soviet intelligentsia, to the entire Soviet people!

Honour and glory to the Soviet scientists, engineers, and technicians who made the spaceship!

Honour and glory to Comrade Yuri Alexeyevich Gagarin, the first cosmonaut and pioneer of space exploration!

We Soviet people, who are building communism, had the honour of being the first to penetrate into space. We consider the victories won in space exploration to be achievements, not only of our people, but of mankind as a whole. We gladly put them at the service of all peoples for the benefit of the progress, happiness and welfare of all men on earth. We do not use our achievements and discoveries for war, but for the peace and security of the peoples.

Scientific and technological progress affords unlimited opportunities of harnessing natural forces and utilising them for the good of man, which calls above all for safeguarding peace.

On this festive day we again address an appeal for peace to the peoples and governments of all countries.

Let all men, irrespective of race or nation, colour, religion or social distinction, do all in their power to assure a lasting peace throughout the world. Let us put an end to the arms race. Let us effect general and complete disarmament under rigid international control. This would be a decisive contribution to the sacred cause of peace.

The splendid victory won by our country will inspire all Soviet people for further great achievements in communist construction.

Forward to new victories for peace, progress, and the happiness of mankind!

CENTRAL COMMITTEE
OF THE COMMUNIST PARTY
OF THE SOVIET UNION
PRESIDIUM OF THE SUPREME SOVIET
OF THE U.S.S.R.
COUNCIL OF MINISTERS
OF THE UNION OF SOVIET
SOCIALIST REPUBLICS

The Kremlin, Moscow,
April 12, 1961

STATEMENT MADE BY Y. A. GAGARIN BEFORE THE TAKE-OFF

Before starting on his space flight in the satellite ship *Vostok* Y. A. Gagarin made the following statement to the press and radio:

"Dear friends, both known and unknown to me; fellow-countrymen, men and women of all lands and continents,

"In a few minutes a mighty spaceship will take me into the far-away expanses of the Universe. What can I say to you in these last minutes before the start? I see my whole past life as one wonderful moment. Everything I have experienced and done till now has been in preparation for this moment. You must realise that it is hard to express my feelings now that the test for which we have been training ardently and long is at hand. I don't have to tell you what I felt when it was suggested that I should make this flight, the first in history. Was it joy? No, it was something more than that. Pride? No, it was not just pride. I felt very happy—to be the first in space, to engage in an unprecedented duel with Nature—could one dream of anything greater than that?

"But then I thought of the tremendous responsibility of being the first to accomplish what generations of people had dreamed of, the first to show man the way into space.... Can you think of a task more difficult than the one assigned to me. It is not responsibility to a single person, or dozens of people, or even a collective. It is responsibility to all Soviet people, to all mankind, to its present and its future. And if I am nevertheless venturing on this flight, it is because I am a Communist, because I draw strength from unexampled exploits performed by my compatriots, Soviet men and women. I know that I shall muster all my will power the better to do the job. Realising its importance, I will do all I can to carry out the assignment of the Communist Party and the Soviet people.

"Am I happy to be starting on a space flight? Of course I am. In all times and all eras man's greatest joy has been to take part in new discoveries.

"I would like to dedicate this first space flight to the people of communism, a society which our Soviet people are already entering, and which, I am confident, all men on earth will enter.

"It is a matter of minutes now before the start. I say to you good-bye, dear friends, just as people say to each other when setting out on a long journey. I would like very much to embrace you all—people known and unknown to me, close friends and strangers alike.

"See you soon!"

SPACEPORT

From V. GOLTSEV and N. DRACHINSKY,
Izvestia Special Correspondents

April 12, 1961

On the morning of April 12, 1961, the sun rose punctually, as always, and sent its first rays through the light curtains of the quiet room where calmly slept a man whose name was to become known to the whole world in a few short hours.

A doctor looked into the room. "He's still sleeping."

Everybody was in a state of excitement—the doctors, the engineers, the scientists. Everybody but this man who was to make the first space flight in history. He slept calmly on. The previous evening, in keeping with the regimen, the doctors had told Yuri Gagarin to sleep ten hours. Everybody was surprised at how quickly he fell asleep. It was as though a fishing trip or a basketball game awaited him the next day instead of a flight into space.

The doctor entered the quiet room. "It's time to get up, Yuri," he said.

Yuri opened his eyes and smiled. He was refreshed, cheerful, buoyant. Springing out of bed, he did his usual morning setting-up exercises.

Men in white smocks came in and helped him to dress.

Getting into the spaceman's complex gear is not a simple matter. The inventive minds of scientists and the skilled hands of craftsmen had prepared an outfit in which Yuri would be warm and comfortable, and his body reliably protected against all the forces he would be subjected to in space.

The men in white checked every fastening and strap of the spacesuit and the helmet.

The preparations over, the spaceman got into a bus, together with his comrades, and rode across the spaceport to the huge spaceship.

A new word, *spaceport*, is now coming into our language. It means a large and complex system of facilities operated by skilled personnel. This is where spaceships are prepared for launching, and where they take off into the starry heavens. All the thoroughly tested equipment was made by Soviet scientists, engineers, and workers, by socialist industry.

The creative genius, skill and ability of Soviet people guaranteed the successful launching of the *Vostok* spaceship with the first space pilot on board.

Engineers, scientists, and workers had already gathered at the foot of the gigantic spaceship. Many of them had known Yuri Gagarin for a long time. There were the last words of parting, embraces, kisses.

This was not the first time the close-knit staff of the spaceport was sending up a huge ship from the launching pad. But this was an unusual day: the first man was setting out into space.

The specialists painstakingly checked everything, making sure the complex apparatus functioned flawlessly.

Yuri was in high spirits. He joked with his comrades, shook hands with them, and thanked them for their best wishes of success.

Now everything was ready. Yuri Gagarin entered the lift that was now to take him up to his cabin at the top of the colossal rocket.

All eyes followed the lift as it rose. "Good luck!" "Happy landing!" the people down below shouted.

Words cannot convey the emotions of all present at the spaceport in those historic moments when they saw off a Soviet man on a flight that was to open up a new era, the era of space travel.

The lift came to a stop. Yuri Gagarin paused for an instant. A last wave to the friends and comrades down below. Then he stepped inside the spaceship.

A few seconds later the command was given. The gigantic ship rose up out of a fiery cloud towards the stars. The events that followed are now known the world over.

REPORT FROM THE LANDING AREA OF THE *VOSTOK*

By Georgi OSTROUMOV,
Izvestia Special Correspondent

April 12, 1961

I arrived at landing site headquarters this morning before news of the launching was announced over the radio.

There were two big maps in the office. A red line marking the spaceship's route ran across one of them. Loudspeakers and telephones lined the desks. The air was fresh—no smoking allowed here. The tiny lamps flashed on and off on the panels. The specialists took up their posts. Everybody was pleased with the sunny, almost windless weather. That made work easier for them, and, most important, easier for the space pilot to land.

The readiness of all the men and machines was checked. The airfield reported the planes and helicopters were ready to take off at an instant's notice. Konstantin Terentyevich, leader of the group, picked up the telephone receiver. There was now a set, concentrated expression on his energetic, large-featured, usually smiling face. As he put down the receiver he was smiling again.

"The launching was successful," he said. "The ship isn't in orbit yet, but the pilot has already sent down a message. He says he sees the earth shrouded in a haze."

A minute or so later the TASS announcement of the start of the historic flight came over the radio.

It took the spaceship just under an hour and a half to carry the first man round the globe. Magellan's voyage round the world took three years. Speeds on the space roads are different.

An hour and a half in a world no human being had ever been in before! It fell to the eyes of a Soviet space pilot to see the true picture of the sky—its real, unfamiliar colour, the primordial brightness of the stars and the sun. He is the first to be able to say: I actually saw that the earth is round. He is the first to have ceased feeling the weight of his body, for an hour and a half, and to be able to answer hundreds, perhaps thousands, of questions in which science today is eagerly interested.

As the space pilot, a son of the Land of Soviets, continued his flight we at the landing site headquarters, along with everybody else in the world, were interested mainly in one thing: how did he feel up there? The radio gave us the answer. As Gagarin sped over South America he reported: "Flight proceeding normally, am feeling fine."

There was good reason for surprise. The ship hurtled through almost airless space at a speed of nearly eight kilometres per second, at an altitude of about 300 kilometres, in cold so intense it could hardly be measured—yet the flight proceeded normally.

Yesterday evening the *Pravda* correspondent and I interviewed Dr. Vitaly Valovich, a member of the group that was to come down by parachute at Gagarin's landing place.

"The painstaking training, the excellent, thoroughly tested equipment, and the experience which Soviet scientists and designers have gained in launching spaceships carrying animals provide a reliable guarantee of safety," he said.

Earth lay far below the spaceship. But the pilot did not feel cut off. His native land broadcast music: from Moscow, songs about the Soviet capital, from Khabarovsk, the song "Amur Waves",

One of the office telephones rang out. "Yuri Gagarin's here!" an excited voice said.

This was a call from a man who had already shaken the intrepid spaceman's hand. Gagarin had made an excellent landing. He had not waited for a helicopter but had gone forward to meet the people who had seen him coming down.

The flight that will be talked about for years, for centuries, was over. The Soviet man was back from his heroic journey through space.

A helicopter brought Yuri Gagarin to the nearest town where he spoke by telephone with N. S. Khrushchov. The First Secretary of the Communist Party of the Soviet Union heartily congratulated Communist Yuri Gagarin, conqueror of space, a son of the great socialist homeland.

* * *

Sports Commissar I. G. Borisenko arrived together with Gagarin. He had registered three world records set up by Gagarin during his amazing flight: an altitude record for manned space flight, a duration record, and a payload record.

I have just seen Yuri Gagarin. He came out of the plane smiling the smile of a truly happy man. He wore a sky-blue flying suit and helmet. People rushed up to embrace and kiss and congratulate him. He hugged one of the men, obviously an old friend, so hard it looked for a moment as though they were wrestling. Yuri was in the best of spirits.

"Hearty congratulations from the readers of *Izvestia*," I said.

"Pass on my sincere greetings to them," he replied. The eyes of the first space pilot shone. The glow of the stars still seemed to be reflected in them.

To Major Yuri Alexeyevich Gagarin, the Soviet Cosmonaut Who Was the First in the World to Accomplish a Space Flight

Dear Yuri Alexeyevich,

I take great pleasure in heartily congratulating you on your outstanding heroic feat, the first space flight in the satellite ship *Vostok*.

All Soviet people applaud your splendid feat, which will be remembered through the ages as an example of courage, daring, and heroism in the service of mankind.

The flight you have accomplished opens a new page in world history and the conquest of space; it fills the hearts of Soviet people with great joy and makes them proud of their socialist country.

From the bottom of my heart I congratulate you on your safe return to your native soil after your space trip. I embrace you.

Till we meet soon in Moscow.

N. KHRUSHCHOV

April 12, 1961

"A FEAT THAT WILL LIVE THROUGH THE AGES"
CONVERSATION BETWEEN N. S. KHRUSHCHOV AND Y. A. GAGARIN, THE FIRST SPACE PILOT

N. S. Khrushchov, First Secretary of the C.C. C.P.S.U. and Chairman of the Council of Ministers of the U.S.S.R., staying in the vicinity of Sochi, on April 12 followed with keen attention the preparations for and launching of the satellite ship *Vostok*, and the flight of the pilot Yuri Alexeyevich Gagarin, the first cosmonaut in the world, Soviet citizen, Communist, pioneer space explorer.

Shortly after it was reported that the first space flight had been safely completed and Y. A. Gagarin had landed in the assigned area, a telephone conversation took place between Premier Khrushchov and the first space pilot, Gagarin. It was at 1300 hours Moscow Time. Premier Khrushchov was informed that Gagarin would like to speak to him.

"I shall be very happy to talk with Comrade Gagarin," said Premier Khrushchov.

Taking the receiver, he said, "I'm glad to hear your voice, dear Yuri Alexeyevich."

GAGARIN. I have just received your telegram of greetings congratulating me on the successful completion of the first space trip in the world. I wish to give you my heartfelt thanks for your congratulations, Nikita Sergeyevich. I am happy to report that the first space flight has been successfully completed.

KHRUSHCHOV. I heartily greet and congratulate you, dear Yuri Alexeyevich. You were the first in the world to accomplish a space flight. By your feat you have brought fame to our country, have shown courage and heroism in carrying out an important assignment. Your feat has made you an immortal man, for you are the first man to have penetrated into space.

Tell me, Yuri Alexeyevich: How did you feel in flight? How was that first space flight?

GAGARIN. I felt all right. The flight was most successful. All the equipment of the spaceship functioned with precision. During the flight I saw the earth from a great altitude. I could see seas, mountains, big cities, rivers, and forests.

KHRUSHCHOV. So you may be said to have felt fine?

GAGARIN. You put it correctly, Nikita Sergeyevich— I felt quite at home in the spaceship. I thank you again for your congratulations and greetings on the occasion of the successful completion of the flight.

KHRUSHCHOV. I am glad to hear your voice and to greet you. I shall be glad to see you in Moscow. We shall celebrate, together with you and with all Soviet people, that great feat in space exploration. Let the world look on and see the things our country is capable of doing, the things our great people and our Soviet science can do.

GAGARIN. Now let other countries try and overtake us.

KHRUSHCHOV. Exactly! I am glad that your voice sounds cheerful and confident, and that you are in such wonderful spirits. You are right—let the capitalist countries now try to overtake our country, which has broken a path into space by sending up the world's first space pilot. We are all proud of this great victory.

Anastas Ivanovich Mikoyan is here with me. He asks me to give you his heartfelt congratulations and regards.

GAGARIN. Please convey my thanks and best wishes to Anastas Ivanovich.

KHRUSHCHOV. Tell me, Yuri Alexeyevich, do you have a wife and children?

GAGARIN. I have a wife—her name is Valentina Ivanovna—and two daughters, Lena and Galya.

KHRUSHCHOV. Did your wife know you were going to fly into space?

GAGARIN. Yes, she did, Nikita Sergeyevich.

KHRUSHCHOV. Give my best regards to your wife and your children. May your daughters grow up in the proud knowledge that their father has accomplished so great a feat in behalf of our Soviet country.

GAGARIN. Thank you, Nikita Sergeyevich, I will convey your regards. I shall always remember your heartfelt words.

KHRUSHCHOV. Are your parents, your mother and father, living? Where are they now? What do they do?

GAGARIN. My father and mother are living, they live in Smolensk Region.

KHRUSHCHOV. Give your father and mother my hearty congratulations. They have good reason to be proud of their son who has accomplished so great a feat.

GAGARIN. Thank you very much, Nikita Sergeyevich, I will convey your words to my father and mother. They will be glad to hear them and will be very grateful to you and to our Party and the Soviet Government.

KHRUSHCHOV. It is not only your parents—the whole of our Soviet country is proud of your great feat, Yuri Alexeyevich. You have accomplished a feat that will live through the ages.

Once again I congratulate you with all my heart on the

successful completion of the first space flight. See you soon in Moscow. My best wishes.

GAGARIN. Thank you, Nikita Sergeyevich. I wish to thank you again and our Communist Party and the Soviet Government for the confidence placed in me, and I assure you I am willing to carry out any further assignment of our Soviet country. Good-bye, dear Nikita Sergeyevich!

SPACE PILOT SPEAKING

From G. OSTROUMOV, *Izvestia* Special Correspondent

Landing Area of the Spaceship *Vostok*
April 13, 1961

I met Yuri Alexeyevich Gagarin, the first space pilot, this morning, April 13. He is in high spirits, hale and hearty. A wonderful smile illumines his face. Today he is no longer in his space flying suit, but is wearing his officer's uniform. Incidentally, one more detail has been added to it—a badge showing Yuri Gagarin to be a first-class flyer. He was awarded the title this morning.

We begin our interview. The space pilot answers questions by *Izvestia* and *Pravda* correspondents quickly and to the point. Every now and then dimples appear on his cheeks. He appreciates the curiosity with which he is pressed for the details of what he saw and experienced during the one and a half hours he spent outside the earth.

He searches for words to describe his sensations as precisely as possible. Now and again he repeats himself so that his listeners may get a better idea of what he is saying. The only person who has seen a different earth, he wants his words to convey to all the 3,000 million inhabitants of our planet the picture that opens before one's eyes beyond the atmosphere, and to give them at least an inkling of the emotions experienced by the pilot of the marvellous spaceship.

Here is the sum and substance of our interview.

Question: What did you feel before boarding the spaceship?

Answer: I boarded the spaceship with a feeling of great satisfaction. I was happy and proud that the flight in space was to be made by me. At the same time I had a sense of particular responsibility for the flight into space where there is so much that is unknown. I was proud for our people who were able to build ships powerful enough to lift man into outer space.

Question: What did you feel and what did you think about during the flight?

Answer: All my thoughts and feelings were concentrated on carrying out the flight programme. I wanted to carry out my assignment most thoroughly and to the best of my ability. There was a lot of work; the flight was all work.

Question: How did you feel when the sense of gravity disappeared during the take-off and when it reappeared?

Answer: When weightlessness came I felt fine. Everything was easier to do. And this stands to reason. My legs and arms did not weigh anything. The objects floated in the cabin. Nor did I keep my seat as I had until then, but was suspended in the air. During the state of weightlessness I ate and drank, and everything went on as it does on earth. I even worked in this state; I wrote down my observations. My handwriting was the same, although my hand did not weigh anything. I had to hold on to the notepad, though, or it would have floated away. I maintained radio communication in different channels and used a telegraph key. I saw that weightlessness does not in any way affect one's capacity for work. The transition from weightlessness to gravitation, to the appearance of gravity was smooth. The arms and legs felt as they did during the state of weightlessness with the only difference that they now had weight. I, too, was no longer suspended above my seat but resumed it.

Question: How do the day and night sides of the earth appear from a high altitude? What do the sky, sun, moon and stars look like?

Answer: The day side of the earth can very well be seen from a great altitude. I could clearly discern the continental coastlines, islands, large rivers and water reservoirs, and relief. When I flew over our country I distinctly saw the large squares of collective-farm fields and could easily tell ploughed land from meadows. Before my space flight I never rose higher than 15,000 metres. Of course, visibility from a spaceship is poorer than it is from a plane, but, nevertheless, one can see things very well. During the space flight I saw the spherical form of the earth for the first time with my own eyes. You get the same picture when you look at the horizon. It should be noted that the horizon presents a very unique and unusually beautiful sight. One can observe an uncommonly colourful transition from the light surface of the earth to the perfectly black sky with stars shining in it. This transition is very fine, it is like a film surrounding the earth. The film is a delicate blue, and the transition from blue to black is unusually smooth and beautiful. It is quite difficult to describe. But when I emerged from the earth's shadow the horizon had changed. It had a bright orange strip which changed to blue and then to deep black again. I did not see the moon. In space the sun shines many times as bright as here on earth. The stars can be seen very well: they are bright and clear-cut. The entire firmament appears in much sharper relief than we see it from the earth.

Question: Did you feel lonely out in space?

Answer: No, I certainly did not feel lonely at all, for I knew that my friends, in fact all the Soviet people, were watching my space flight. I was sure that the Party and the Government were ready to help me should I find myself in straits.

Question: Where were you when the first Soviet sputnik was launched, what were you doing and did it occur to you that you may be the first space flyer?

Answer: When the first Soviet artificial satellite of the earth was launched I was finishing the Orenburg flying school. My schoolmates and I learned about this event after our flying exercises in MIGs that very same day. We were very proud of the great achievement of Soviet science and engineering. It was clear that before long a human being would fly into space. And yet I thought it would probably be another ten years. However, it took less than four years. Of course, at that time I, too, wanted to fly to outer space, but it never occurred to me that I may be precisely the one to fly the first spaceship.

Question: When you were a schoolboy what subjects did you like best?

Answer: I went through six grades of the Gzhatsk secondary school. Then I attended the Lyubertsy trade school and after that a Saratov industrial school. All through my school years I was particularly fond of two subjects—physics and mathematics.

Question: When did you hear the name of Tsiolkovsky for the first time?

Answer: I heard the name of Tsiolkovsky for the first time when I was still a schoolboy. When I attended the trade and industrial schools I learned to love this name. I studied Tsiolkovsky's works. I must say that in his book *Outside the Earth* Tsiolkovsky very clearly foresaw all that I chanced to see with my own eyes during my flight. Tsiolkovsky, like no one else, had a clear idea of the world as it appears to a person in outer space.

Question: Who is your favourite literary character? Favourite writer?

Answer: I like many writers, both Soviet and classic. I am very fond of reading Chekhov, Tolstoi, Pushkin and Polevoi. My most favourite literary character, the one I learned to love when I was still a boy, is the main character in Boris Polevoi's book *A Story about a Real Man*. I regret very much that I have never had the chance of meeting Maresyev. I also read Jules Verne. Of course, his books are very interesting, but life proved things to be different. Ivan Yefremov's novel *Andromeda* is good and I liked it. However, since I have been in outer space I realise that not everything in this book is true to fact. But it is useful just the same.

Question: Could you, judging by the way you felt, have stayed in space longer?

Answer: In the spaceship I could have stayed much longer, but the duration of my flight was determined by a programme beforehand. I could work very well on board the spaceship, I felt well and was in high spirits. I could have kept flying in space as long as was necessary.

Question: What was your first feeling when you touched ground?

Answer: I can hardly describe my feelings when I stepped again on our Soviet soil. To begin with, I was happy to have carried out my assignment. I was just happy about everything. As I was descending I kept singing the song "My country hears, my country knows...".

Question: What did you think when you were asked to start training for a flight into space?

Answer: Personally, I was very anxious to fly to outer space. I wanted to become a space pilot. When I was entrusted with the flight I began to train for it. As you see, my wish has been fulfilled.

Question: Do you go in for sports? What sport do you like best?

Yuri GAGARIN, pilot of the *Vostok* spaceship

In the bus on the way to the launching site

Yuri GAGARIN before launching

Homebound after the descent

Y. A. GAGARIN reports on his successful landing to N. S. KHRUSHCHOV

School children attending a lecture at the Moscow Planetarium cheer the news of space pilot Yuri GAGARIN's safe return

Yuri GAGARIN is given a gala welcome at Moscow's Vnukovo Airport

Yuri GAGARIN and Premier KHRUSHCHOV on the way
to Red Square from Vnukovo Airport

Red Square, Moscow, April 14, 1961. Y. A. GAGARIN, N. S. KHRUSHCHOV, F. R. KOZLOV and L. I. BREZHNEV on the Mausoleum reviewing stand

Thousands of Muscovites cheer the space hero in Red Square

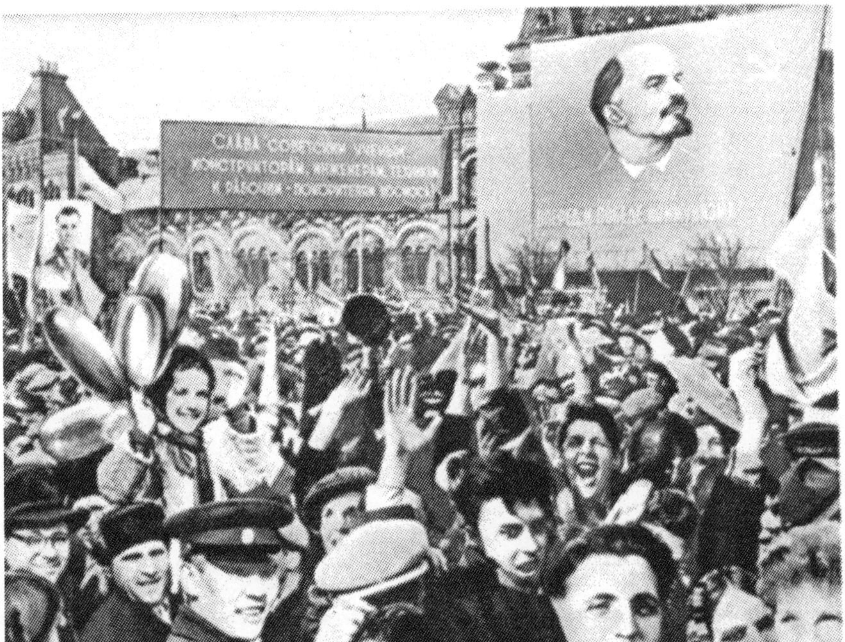

The space pilot with wife Valentina and daughter Yelena

Yuri GAGARIN addresses a press conference on the first manned space flight

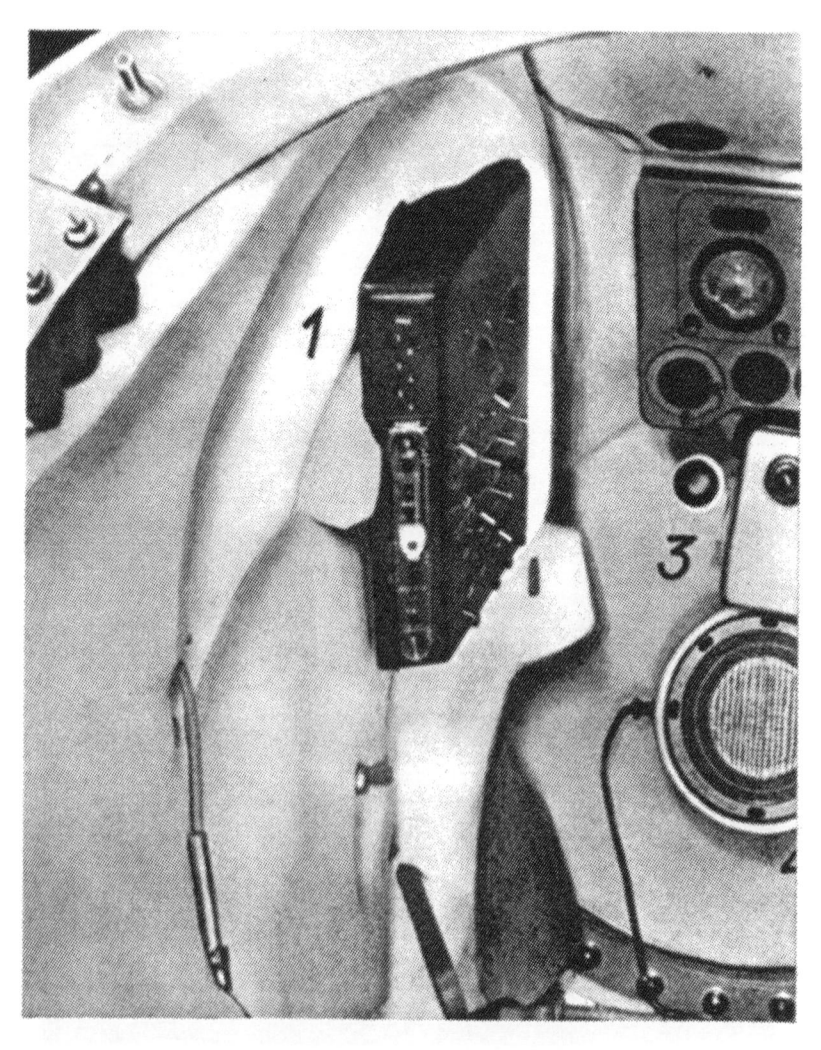

The inside view of the spaceship *Vostok*: 1—control panel; 2—instrument board with a globe; 3—TV camera; 4—porthole with an optical orientor; 5—spaceship orientation control stick; 6—WT set; 7—food container

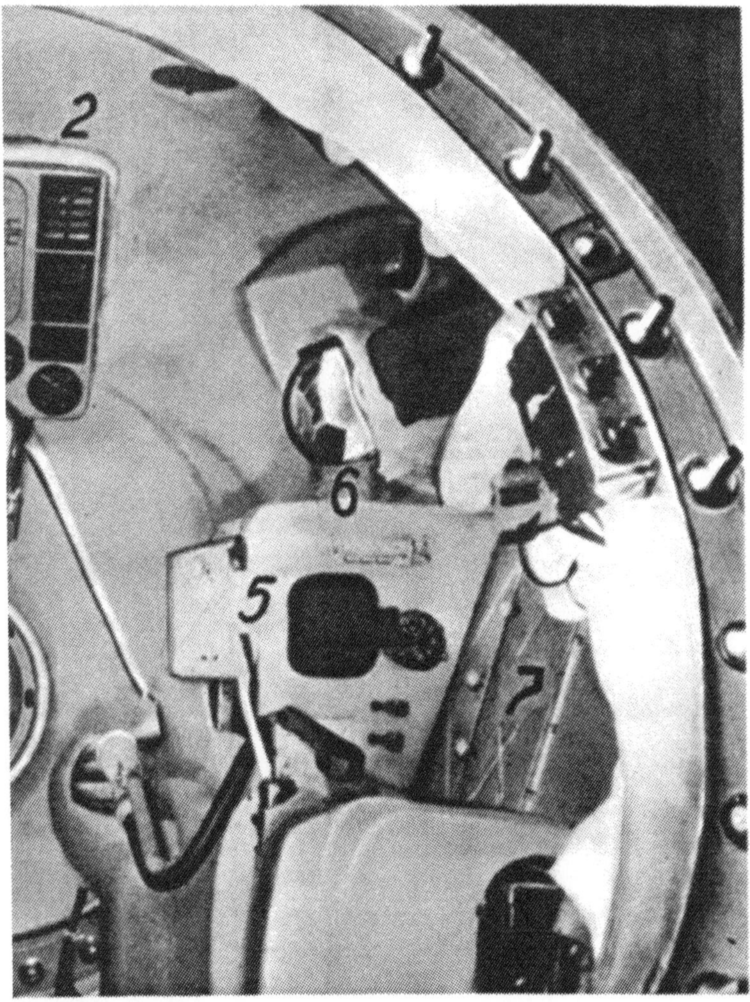
2
6
5
7

Ascent into orbit

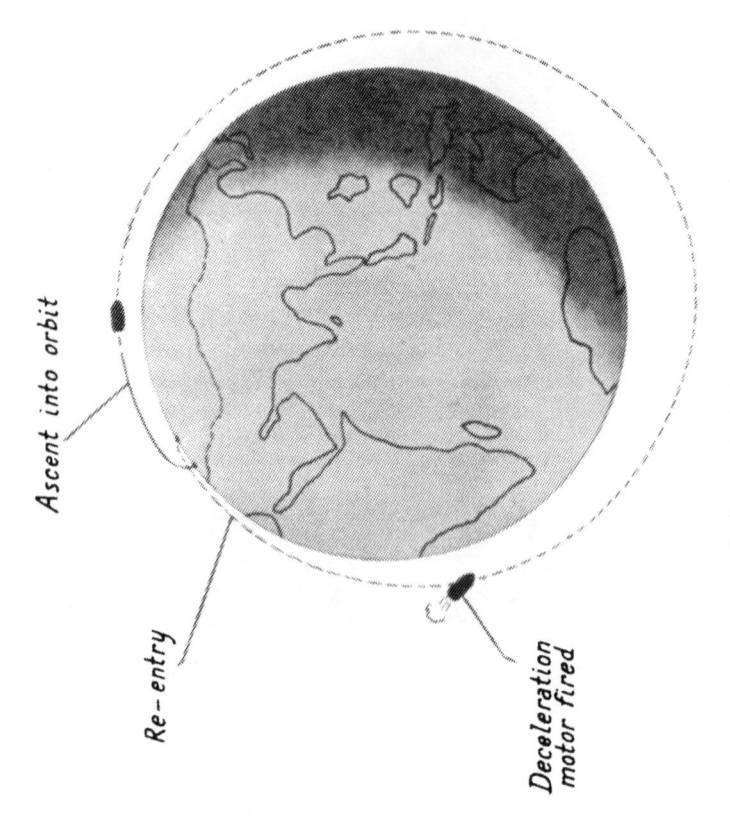

Re-entry

Deceleration motor fired

Flight path diagram of the spaceship *Vostok*

Answer: I like sports, basketball in particular. Besides, I like to ski, skate, and play badminton. Badminton is a very good and healthy game.

Question: What is your favourite occupation?

Answer: I like flying best. I used to fly planes, but I like my space flight best of all. Can my first flight in an aircraft be compared with my flight of yesterday? It is hard to compare them. The former was a flight in a winged craft, the latter—in a wingless one. The plane flew at a speed of 150 km/hr., the spaceship—28,000 km/hr. The former flew at an altitude of 1,500 metres, the latter—300 km.

Question: What made you particularly happy when you returned to the earth?

Answer: When I returned to the earth I rejoiced at the warm reception I was accorded by the Soviet people. I was moved to tears by the telegram I received from Nikita Sergeyevich Khrushchov. I was moved by his concern, attention and cordiality. I felt the happiest when I spoke to Khrushchov and Brezhnev on the telephone. I am sincerely and filially grateful to Nikita Sergeyevich for his encouragement and concern.

Question: According to foreign press reports, the U.S.A. also intend to send a man into space. What can you say about it?

Answer: Our Party and Government have raised the question of peaceful utilisation of space and peaceful competition. We shall, of course, rejoice at the achievements of American spacemen when they fly into space. There is enough room for everybody there. But space must be used only for peaceful and not military purposes. The American space flyers will have to catch up with us. We shall hail their accomplishments but shall always try to keep ahead of them.

Question: What event was the most significant in your life prior to your space flight?

Answer: In the summer of 1960 I joined the Party. This was the greatest and most memorable event in my life prior to my space flight. I dedicate my flight to our Party, our Government, the Twenty-Second Congress of the Party, all our people who are marching in the forefront of humanity and are building a new society.

Question: What are your plans for the future? Would you be willing to undertake another flight?

Answer: My plans for the future are as follows: I want to devote my life, my work, all my thoughts and feelings to the new science of the conquest of space. I should like to go to Venus to see what is under that planet's veil of clouds; I should like to visit Mars, to see for myself whether there are any canals on it. The moon is not so distant a neighbour of ours and I don't think it will be very long before we can fly to the moon and land on it.

Question: Do you know that after the launching of the first sputnik thousands of letters were received from people who asked to be sent into space? Have you read these letters?

Answer: Yes, I have read them. All those letters are very sincere. Of course, I feel sorry that these people have not yet been able to make the flight, but I am sure that the time will come when people will get trade-union spaceship holiday accommodations.

Question: What do you want us to tell your parents and fellow-townsmen?

Answer: Please give my parents and fellow-townsmen my regards, and best wishes for success in life and work.

The question period came to an end. Of course, we could go on asking more and more questions and listen to answers to them, but the space flyer does not have very much time. Besides, he will yet have a good deal to tell us.

Before taking leave I asked him for an autograph for my newspaper.

"This Feat Is an Embodiment of the Genius of the Soviet People and the Great Might of Socialism"

GLORY TO THE SOVIET SCIENTISTS, DESIGNERS, ENGINEERS, TECHNICIANS, AND WORKERS CONQUERING SPACE!

TO ALL THE SCIENTISTS, ENGINEERS, TECHNICIANS, AND WORKERS, TO ALL COLLECTIVES AND ORGANISATIONS, THAT TOOK PART IN THE SUCCESSFUL ACCOMPLISHMENT OF THE WORLD'S FIRST MANNED SPACE FLIGHT IN THE SATELLITE SHIP *VOSTOK* TO COMRADE YURI ALEXEYEVICH GAGARIN, THE FIRST SOVIET COSMONAUT

Dear comrades,

Fellow-compatriots,

The peoples of our country have witnessed a joyous, stirring event. On April 12, 1961 our country, the Union of Soviet Socialist Republics, for the first time in the history of mankind successfully sent a man into space in the satellite ship *Vostok*.

The flight of a Soviet man into space is a tremendous achievement of the creative genius of our people, and a result of the free, inspired effort of Soviet people, who are building communism. The dream of outstanding men of Russian and world science and technology, a dream to which

Konstantin Eduardovich Tsiolkovsky, an outstanding son of our people, devoted his life, has now materialised, becoming a living reality of our heroic times. It is a great, outstanding contribution made by the Soviet people to the treasury of world science and culture. Mankind will accept this invaluable service of the Soviet Union with gratitude. The heroic flight of a Soviet man into space has ushered in a new era in the history of the earth. An age-long dream of mankind has come true.

On behalf of our glorious Communist Party, the Soviet Government and all peoples of the Soviet Union, the Central Committee of the Communist Party of the Soviet Union, the Presidium of the Supreme Soviet of the U.S.S.R. and the Council of Ministers of the U.S.S.R. extend their warm congratulations on the great victory of human genius and labour to all the scientists, designers, technicians and workers, to all the collectives and organisations, that took part in the successful accomplishment of the world's first manned space flight.

We heartily greet and congratulate you, our dear Comrade Yuri Alexeyevich Gagarin, on the occasion of a supreme feat, the first space flight.

Our free, talented and industrious people, whom the Communist Party, headed by Vladimir Ilyich Lenin, the great leader and teacher of the working people of the world, roused in October 1917 for conscious history-making, are today demonstrating to the whole world the immense advantages of the new, socialist system in all spheres of the life of society.

The manned space flight is a result of the successful realisation of the vast programme of full-scale communist construction, of the unflagging attention which the Communist Party and its Leninist Central Committee and the Soviet Government headed by Nikita Sergeyevich Khrushchov devote to the continuous advancement of science, technology and culture and to the good of the Soviet people.

Less than four years had passed between the launching of the world's first man-made earth satellite by the Soviet Union and the successful manned space flight.

Soviet scientists, engineers, technicians, and workers have by their devoted labour opened for man's genius the way into the depths of outer space. They have done so for the benefit of peace on earth, for the happiness of all peoples.

Man's first space flight will become a new source of inspiration and daring endeavour to all Soviet people for the benefit of further progress and world peace.

Glory to the Soviet scientists, designers, engineers, technicians, and workers conquering space!

Glory to our people, a people of creators and victors, who under the leadership of the Communist Party are blazing the road to a bright future for mankind—communism!

Long live the glorious Communist Party of the Soviet Union, which inspires and organises all the victories of the Soviet people!

Long live communism!

CENTRAL COMMITTEE
OF THE C.P.S.U.
PRESIDIUM OF THE SUPREME
SOVIET OF THE U S.S.R.
COUNCIL OF MINISTERS
OF THE U.S.S.R.

NATION HAILS THE HERO

PIONEER OF THE UNIVERSE GREETED
Report from Vnukovo Airport

Today Vnukovo Airport attracted the hearts and minds of all Soviet people wherever they were and whatever they happened to be doing.

L. I. Brezhnev, N. G. Ignatov, F. R. Kozlov, A. N. Kosygin, O. V. Kuusinen, N. A. Mukhitdinov, N. V. Podgorny, D. S. Polyansky, Y. A. Furtseva, N. M. Shvernik, P. N. Pospelov, D. S. Korotchenko, J. E. Kalnberzin, A. P. Kirilenko, K. T. Mazurov, G. I. Voronov, and V. V. Grishin arrived at the airport soon after midday. Among those who had come to meet the hero were members of the Presidium of the Supreme Soviet of the U.S.S.R. A. A. Andreyev and K. Y. Voroshilov, Deputy Chairmen of the Council of Ministers of the U.S.S.R., ministers of the U.S.S.R. and the R.S.F.S.R., chairmen of State Committees, members of the Central Committee of the C.P.S.U. and candidate-members of the Central Committee of the C.P.S.U., marshals of the Soviet Union, heads of Party organisations and local government bodies of Moscow and Moscow Region, and many others. Representatives of the working people of Moscow gathered, holding flowers and banners with words of greetings.

Members of the diplomatic corps and Soviet and foreign correspondents were at the airport.

Soon the courageous cosmonaut's father, mother and wife came in a car. They were warmly greeted. Nina Petrovna Khrushchova presented a bouquet of flowers to Valentina Gagarina.

The hands of the clock showed 1230 hours. An IL-18 airliner landed on the far runway and slowly taxied towards the airport building. This was a special plane bringing N. S. Khrushchov from the vicinity of Sochi to meet the hero. N. S. Khrushchov and A. I. Mikoyan and V. P. Mzhavanadze, who arrived with him, were greeted with applause. Khrushchov waved his hat, greeted and embraced comrades and friends.

L. I. Brezhnev introduced the hero's father, Alexei Ivanovich Gagarin, to N. S. Khrushchov. They embraced. At that very moment, another IL-18, escorted by seven jet fighters, appeared overhead. It carried the man whose name is today on people's lips throughout the world. He is Major Yuri Gagarin, the first space traveller, a citizen of the Soviet Union.

The planes winged off to make a circle of honour over Moscow, and the leaders of the Communist Party and Soviet Government and members of the hero's family went to the reviewing stand. N. S. Khrushchov stopped at the foot of the stand to let Yuri Gagarin's mother, father and wife pass.

A few more minutes of waiting and the aircraft carrying Yuri Gagarin taxied up to the carpeted strip of the tarmac. The door opened and the trim Major Gagarin went down the gangway, and as the band played the Air Force March he strode towards the stand.

All of us, our whole country and the whole world were deeply moved by the simple and significant words of his report:

"Comrade First Secretary of the Central Committee of the Communist Party of the Soviet Union, Chairman of the

Council of Ministers of the Union of Soviet Socialist Republics,

"I am happy to report that the assignment of the Central Committee of the Party and the Soviet Government has been successfully fulfilled.

"*The first space flight in the history of man was successfully made in the Soviet spaceship* Vostok *on April 12. All the instruments and equipment on board worked faultlessly and with precision.*

"I am in excellent health.

"I am ready to carry out any other assignment of our Party and Government.

"Major GAGARIN"

The report concluded, he went up to Khrushchov. The two men cordially shook hands and embraced each other.

"Congratulations, congratulations," Khrushchov said.

The space hero was warmly congratulated by other leaders of the Communist Party and Soviet Government.

His mother clung to him. His father's eyes glistened with tears. Valentina awaited her "turn" to embrace and kiss her husband.

Then Khrushchov introduced Major Gagarin to Y. Tsedenbal, Chairman of the Council of Ministers of the Mongolian People's Republic, and to the members of the diplomatic corps. They walked past representatives of Moscow's working people, who warmly greeted them.

In the meantime, a flower-decked open car was already drawn up at the stand. Khrushchov entered the car together with Yuri and Valentina Gagarin, and the triumphal motorcade drove off heading towards Red Square.

MEETING AND DEMONSTRATION IN RED SQUARE
April 14, 1961

Since morning all the streets of the Soviet capital leading to Red Square were filled with thousands of Muscovites moving in festive columns to extend a hearty welcome to space hero Yuri Alexeyevich Gagarin.

Red Square, where Soviet people celebrate their holidays and which has witnessed many major events in the history of our country, was more beautiful than ever.

Opposite the Mausoleum, across the façade of the GUM Department Store, hung a portrait of Lenin against the background of a huge red flag with the words beneath it: "Forward, to the triumph of communism!" Big letters formed the slogan, "Long Live the Glorious Communist Party Created by Lenin!"

A huge panel with a portrait of Lenin and next to it a picture of a spaceship and its pilot Yuri Gagarin, adorned the Museum of History. There were scarlet banners with the words: "Glory to Soviet Scientists, Designers, Engineers, Technicians, and Workers—the Conquerors of Outer Space!" and "Honour and Glory to Comrade Yuri Alexeyevich Gagarin, Pioneer Space Explorer!"

Adorned with the coats-of-arms of the fraternal Union Republics, the stands along the Kremlin wall were filled with guests. Among them were workers of factories and mills, foremost collective farmers, scientists and workers in culture, statesmen and public figures, deputies to the Supreme Soviet of the U.S.S.R., the R.S.F.S.R. and the other Union Republics, marshals of the Soviet Union, and generals and officers of the Soviet Army. Also present were members of the diplomatic corps and foreign guests.

From end to end the square was alive with people. Above their heads were a sea of banners, portraits of Lenin and the leaders of the Party and Government, and portraits of Yuri Gagarin.

The square resounded with cheers as the people greeted Comrades *L. I. Brezhnev, N. G. Ignatov, F. R. Kozlov, A. N. Kosygin, O. V. Kuusinen, A. I. Mikoyan, N. A. Mukhitdinov, N. V. Podgorny, D. S. Polyansky, Y. A. Furtseva, N. S. Khrushchov, N. M. Shvernik, P. N. Pospelov, D. S. Korotchenko, J. E. Kalnberzin, A. P. Kirilenko, K. T. Mazurov, V. P. Mzhavanadze, G. I. Voronov, V. V. Grishin,* and member of the Presidium of the Supreme Soviet of the U.S.S.R. *K. Y. Voroshilov,* as they appeared on the Mausoleum reviewing stand. *Y. Tsedenbal,* Chairman of the Council of Ministers of the Mongolian People's Republic, also appeared on the Mausoleum reviewing stand.

With the leaders of the Party and Government were space hero Major Yuri Gagarin and his wife and parents.

On behalf of the Central Committee of the Communist Party of the Soviet Union, the Council of Ministers of the U.S.S.R. and the Presidium of the Supreme Soviet of the U.S.S.R., the meeting of the working people of Moscow, dedicated to the great epoch-making achievement of the Soviet people—the successful accomplishment of history's first space flight by Yuri Alexeyevich Gagarin—was opened by F. R. Kozlov, member of the Presidium of the Central Committee of the C.P.S.U. and Secretary of the Central Committee of the C.P.S.U.

F. R. Kozlov gave the floor to the world's first spaceman Yuri Gagarin, who was greeted with stormy applause.

SPEECH BY Y. A. GAGARIN

My dear countrymen,
Dear Nikita Sergeyevich,
Comrades Party and Government leaders,

To begin with, allow me to express my sincere gratitude to the Central Committee of my Party, to the Soviet Government, all the Soviet people, and to you, Nikita Sergeyevich, for the great trust shown me, an ordinary Soviet airman, by giving me the responsible assignment of making the first flight into space.

When I was starting out into space I was thinking of our Leninist Party, of our socialist homeland.

Love of our glorious Party, of our Soviet homeland, of our heroic and industrious people inspired me and gave me the strength to perform this feat. (*Stormy applause.*)

It is the genius, the heroic labour of our people that created *Vostok*, the finest spaceship in the world, and its very clever, very reliable equipment. From start to finish I had no doubt whatsoever in the successful culmination of the space flight.

I want to thank our scientists, engineers, technicians, and all Soviet workers from the bottom of my heart for building a ship in which one can confidently explore the secrets of outer space. Allow me also to thank all the comrades and all the people who prepared me for the space flight. (*Applause.*)

I am sure that all my friends, my fellow space pilots are also ready at any time to fly round our planet. (*Prolonged applause.*)

It is quite certain that we will fly more distant routes as well in our Soviet spaceships. I am happy beyond all bounds that my beloved country has been the first in the world to perform this flight, the first in the world to penetrate into space. The first plane, the first sputnik, the first spaceship, and the first space flight—those are the milestones of my country's splendid advance towards unravelling the secrets of Nature. (*Applause.*) It is our dear Communist Party that has led and confidently leads our people to that goal. (*Prolonged applause.*)

Throughout my life and study at the trade school, the industrial school, the flying club, and the flying school I have been conscious of the constant concern of the Party, a son of which I am. (*Applause.*)

Dear comrades, I should like to make a special mention of the immense fatherly concern for us, Soviet people, shown by Nikita Sergeyevich Khrushchov. It was you, Nikita Sergeyevich, who was the first to congratulate me warmly on the success of the flight a few minutes after I landed, after I returned from outer space to our native earth. (*Applause.*)

Thank you very much, dear Nikita Sergeyevich, on my own behalf and on behalf of my fellow astronauts! (*Stormy applause.*) We have dedicated this first flight into space to the Twenty-Second Congress of the Communist Party of the Soviet Union. (*Prolonged applause.*)

Thank you heartily, dear people of Moscow, for this warm reception. (*Stormy applause.*) I am sure that under the guidance of the Leninist Party every one of you is ready to perform any feat for the might and prosperity of our beloved homeland, the glory of our country, of our people. (*Stormy applause.*)

Long live our socialist land! (*Stormy applause.*)

Long live our great and powerful Soviet people! (*Stormy applause.*)

Glory to the Communist Party of the Soviet Union and its Leninist Central Committee headed by Nikita Sergeyevich Khrushchov! (*Stormy applause, cheers.*)

A thunderous ovation greeted the next speaker, N. S. Khrushchov, First Secretary of the C.C. C.P.S.U. and Chairman of the Council of Ministers of the U.S.S.R.

The Great Feat Will Go Down in the Ages
SPEECH BY N. S. KHRUSHCHOV

Dear Comrades,
Dear Friends,
Citizens of all the world,

I address you with a sense of great joy and pride. For the first time in history a man from the planet Earth, our Soviet man, broke through into space in a ship created by Soviet scientists, workers, technicians, and engineers, and performed the first unexampled trip to the stars. (*Stormy applause.*)

The satellite spaceship *Vostok* ascended to a height of more than 300 kilometres, girdled the earth and landed safely at a predetermined point in the Soviet Union.

We bid a warm welcome to Yuri Alexeyevich Gagarin, the splendid space pilot, the heroic Soviet man. (*Stormy applause, cheers.*) He has displayed high moral qualities—courage, self-control, and valour. He is the first man to have seen all our planet Earth, which is in perpetual motion, in an hour and a half, taking in its vast oceans and continents at a glance.

Yuri Gagarin is our pioneer of space travel. He has been the first to orbit the globe. If the name of Christopher Columbus, who crossed the Atlantic Ocean and discovered America, has lived down the ages, what can we say about our fine hero, Comrade Gagarin, who penetrated into space, circled the entire globe and landed safely on the earth. His name will be immortal in man's history. (*Stormy applause, cheers.*)

All of us appreciate the world of thoughts and emotions that our first space traveller has brought back with him to the earth. All of us here, on this historic Red Square, share the profound emotion, pride and joy with which we welcome you, our dear friend and comrade. (*Prolonged applause.*)

Allow me to congratulate you heartily and to thank you warmly for your unexampled feat on behalf of the Central Committee of the Communist Party of the Soviet Union, the Soviet Government and all our people. (*Stormy applause.*)

Allow me also to warmly greet and congratulate the scientists, workers, engineers, and technicians who made the rocket ship *Vostok*, to congratulate all the Soviet people who created the conditions for the successful space flight of the manned ship. (*Applause.*)

We are proud of the feat performed by Yuri Gagarin, we admire the scientists, engineers, technicians, and workers who put their intelligence and their heart into the making of this spaceship and its amazing flight. Their splendid performance stems from the devoted effort of all Soviet people—millions of workers, collective farmers, and intellectuals. With this flight we have again demonstrated to the whole world what the genius of a free people can do.

Now that Soviet science and technology have demonstrated a supreme accomplishment of scientific and technical progress, we cannot but look back upon the history of our country. The past years arise involuntarily before the mind's eye of every one of us.

Having wrested power from the tsar, the capitalists, and landlords, we defended it in the flames of civil war, though at times we were unshod and unclothed. How many war strategists there were at the time who prophesied the inevitable defeat of what they called the "ragamuffin armies". But where are these sorry strategists?

When we had our first Communist *subbotniks*, when we laid the foundations of new blast furnaces and built mines, when we flung lofty words to the world, such as *five-year plan*, *industrialisation*, *electrification*, *collectivisation*, *nationwide literacy*, there were many inflated "theoreticians" who forecast that muzhik Russia could not develop into a major industrial power. Where are these sorry prophets today?

We have ever been mindful of our historical past. We have used for the good of the people all the best created by the foremost men of our country. The socialist state furnished an outlet in the broad sphere of Soviet industrial and collective-farm development to the dreams and plans of many scientists, engineers and technicians who in the conditions of tsarist Russia had no chance of applying their minds and hands.

Today, when we stand beside the man who made the first space trip, we cannot but recall the name of Kibalchich, the Russian scientist and revolutionary who dreamt of flights into space and who was executed by the tsarist government. We cannot but recall and pay homage to the memory of Mendeleyev, Zhukovsky, Timiryazev, and Pavlov, and to many other great scientists whose names are associated with the outstanding feats of the Soviet people.

We today recall with special reverence the name of Konstantin Eduardovich Tsiolkovsky, the scientist and visionary, the theoretician of space travel. (*Applause.*)

The dream of conquering outer space is indeed the greatest of all man's great dreams. We are proud that it was Soviet men who made this fabulous dream come true. (*Applause.*)

There is a proud ring to the words Citizen of the Soviet Union. There was a time when people abroad, and some in the country as well, spoke of us with disdain. But even at that time Vladimir Mayakovsky said with pride:

> "*Read,*
> *envy,*
> *I'm*
> *a citizen*
> *of the Soviet Union.*"

(*Prolonged applause.*)

How forcefully these words ring today! How full of profound meaning!

But this pride does not derive from the notion that we deny other nations and countries the performance of some similar feat. We are internationalists. Every Soviet citizen has been brought up in the spirit of socialist patriotism, and at the same time he is ready to share generously his scientific wealth, his technical and cultural knowledge, with anyone who is prepared to live with us in peace and friendship. (*Applause.*)

The Soviet workers, the collective-farm peasantry, the working intelligentsia, are proud that we, the working people of former tsarist Russia, have had the great honour of making the October Socialist Revolution under the guidance of Vladimir Ilyich Lenin, the immortal leader of the working class, and the Communist Party. (*Prolonged applause.*)

That has been a feat unequalled in history. The working class and the people had to show immense courage and bravery, a profound understanding of their goals and tasks, to perform it. The working class did not flinch from any difficulties. It accomplished the greatest of revolutions and took power in a country that was economically backward, almost entirely illiterate, whose people were crushed by tsarism and capitalism.

It was in those conditions, when, it would seem, thought had to be given to ending the war, to healing the bleeding wounds on the whole body of former Russia rather than dream about the lofty accomplishments of our day and the future, that Lenin, a man of vision, spoke with unshakable faith about the inevitable victory of socialism, communism. He took steps to end the imperialist war by means of revolution, by the victory of the working class, the establishment of proletarian dictatorship, the revolutionary emancipation of all the peoples of our country.

Lenin explained insistently and untiringly that a new era will begin in the history of mankind only when people are completely liberated from capitalist slavery, when they gain genuine freedom, when all material and spiritual capacities, and all efforts are concentrated on the good of the working people. (*Applause.*)

The great feat of the Russian working class, the people of our country, who made the October Socialist Revolution under the leadership of the Communist Party, will go down in the ages as an inspiring example of a people's revolutionary endeavour.

Socialism opened up the broadest of vistas for the development of our homeland. In the 43 years of Soviet power the formerly illiterate Russia, of which some spoke with disrespect thinking it to be a barbaric country, has travelled a magnificent road. Now our country has been the first to build the satellite spaceship, the first to break through into outer space. Surely this is a most vivid demonstration of the genuine freedom of the freest of the free peoples of the world— the Soviet people. (*Stormy applause.*)

By creating all the conditions for the take-off and successful landing of the satellite spaceship we have shown what a people can do if it is really free, free politically and economically. It is not the countries where the rich freely exploit those who have no bread and which they call the "free world", but the countries where all working people, all peoples enjoy all the material and spiritual wealth that are really free.

Our conquest of space is a magnificent milestone in the development of mankind. It is a victory that represents a fresh triumph of Lenin's ideas, a confirmation of the Marxist-Leninist teaching. It is a victory of human genius that embodies and graphically reflects all the splendid results of what the peoples of the Soviet Union have achieved in the conditions created by the October Socialist Revolution. It is a feat that marks a new upswing in our country's progressive advance to communism. (*Prolonged applause, cheers.*)

We say with pride and unshakable confidence to the whole world that having successfully accomplished the building of socialism begun in 1917 by the October Revolution, we are marching steadfastly and boldly forward along the path charted by the great Lenin towards communism. We state that there is no force on earth that can turn us from that path. Victory will be ours, and it will be the loftiest and finest of victories. (*Prolonged applause.*) It does not lead to the domination of one group of people over another, the domination of one country over another country or group of countries, of one nation over others, and yields good to all the people of the world. (*Applause.*)

The advance of the peoples to communism, the fine ambition of people to achieve that great goal, cannot be belittled or checked. It is an advance that has gained vast indomitable momentum and there are no obstacles that could check this great process in the development of mankind. The Soviet people, the peoples of the socialist countries, the peoples of all the world, including peoples in countries where they have not yet achieved victory but fight perseveringly for the triumph of progress over exploitation and oppression, will win, will build the bright edifice of communism. And it will be a great boon to mankind, the crowning pinnacle of its continuous development. (*Applause.*)

Comrades, at this hour we hail the scientists of the world, for whom the space flight is a great joy and a great happiness. Soviet science is developing in close association with all world science.

The flight of the spaceship *Vostok* is, so to speak, only the first Soviet swallow in outer space. It soared into the sky in the wake of our many sputniks and spaceships. That is a natural consequence of the gigantic scientific and technological work conducted in our country in the sphere of space exploration.

We shall carry on with this work. Many other Soviet people will fly unexplored routes into space. They will investigate it, further reveal the secrets of Nature and make them serve man, his well-being, and peace.

We stress—make them serve peace! Soviet people do not want the rockets which perform the programme set by man with such amazing precision, to carry deadly loads.

We appeal again to the governments of all the world. Science and technology have advanced so far and are capable, in evil hands, of inflicting such destruction that every effort should be made to achieve disarmament. General and complete disarmament under the strictest international control is the road to lasting peace among nations. (*Stormy applause.*)

When we launched the first sputnik there were short-witted people overseas who did not believe it. Well, there are always short-sighted people around. Now we can touch, so to say, the man who has come back to us straight from the sky! (*Applause.*)

Allow me once again to embrace you, our dear Yuri, and to convey warm greetings through you to your comrades in work and valour. (*Khrushchov embraces Gagarin and kisses him. Stormy applause fills Red Square. Shouts of* "Long live the Communist Party!" *and* "Hail Gagarin!", *cheers.*)

You have added glory to the Union of Soviet Socialist Republics. Your country will never forget your feat and will inscribe your name in its history. (*Applause.*)

We are proud that the world's first space pilot is a Soviet citizen. Yuri Alexeyevich Gagarin grew up and studied in a

Soviet school. He was an active participant in social work, an active Komsomol. He is a Communist, a member of Lenin's great Party! (*Stormy applause.*)

I am pleased to announce that the Presidium of the Supreme Soviet of the U.S.S.R. has awarded you the high title of Hero of the Soviet Union. (*Stormy, prolonged applause, cheers, shouts of* "Hail Hero Gagarin!")

You are also the first to be awarded the fine title of "Pilot-Cosmonaut of the U.S.S.R." (*Stormy applause.*)

A bronze bust of the hero will be erected in Moscow and a medal struck to commemorate the world's first manned flight into space. (*Stormy applause.*)

I heartily congratulate Yuri's parents, Anna Timofeyevna and Alexei Ivanovich Gagarin, on having raised and brought up an excellent son who has added glory to our country with his feat. (*Stormy, prolonged applause.*)

I tender my warm congratulations to Yuri Gagarin's wife, Valentina Ivanovna, a wonderful Soviet woman. She had known that Yuri Gagarin was flying into space and did not try to dissuade him, gave him support, sent off her husband, the father of two small children, with all her heart to perform his great feat. (*Stormy, prolonged applause.*)

After all, no one could guarantee fully that Yuri Gagarin's send-off into space would not be his last. And her courage, her understanding of the meaning of this unexampled flight speak of Valentina Ivanovna's stout heart. (*Prolonged applause.*)

Yes, she is a real Soviet woman. Remember how warmly and affectionately Nekrasov, Pushkin, and our other writers wrote about Russian women. And today all women of the Soviet Union are like those women. Valentina Ivanovna showed great heart, will power, and profound understanding of Soviet patriotism. (*Prolonged applause.*)

Comrades, the peoples of the Soviet Union are celebrating their new victory, a victory of labour, science and reason. It has been won by the peoples of our country through persevering and intensive labour. Soviet people have travelled a great road of struggle for economic development, the progress of technology and science, and have been duly rewarded by winning priority in the launching of the manned satellite spaceship. It is an undying feat, an outstanding accomplishment that will live down the ages as a supreme achievement of mankind. (*Stormy applause.*)

But successes must not blunt our will, our perseverance, our determination to achieve further economic progress and develop science and technology. The tasks of building a solid material and technical basis of communism set by the Twenty-First Congress of the Communist Party are grandiose tasks. They have an immense historic impact. By fulfilling the Seven-Year Plan and thereby achieving a fresh rise of all our economy, science, and technology, we shall produce conditions that will enable us to surpass the economic level of the United States, the most developed capitalist country, and multiply our advantages in scientific and technological development.

By fulfilling the Seven-Year Plan we shall come nearer to crossing the top mark of achievement reached by the capitalist world and break through, as we have now broken through in space, in the development of all our economy, in the satisfaction of the requirements of the people. The material and cultural requirements of Soviet people will be satisfied more fully than the most highly developed countries of the capitalist world can satisfy them.

This is the reason why, comrades, new big successes must not blunt the will, the determination to use all our possibilities in the development of science and technology to best advantage. It is essential to make everything serve the people in order to successfully accomplish the task set by our Party for the development of industry and the country's national economy as a whole.

Especially big tasks confront the Communists and Komsomol members in the countryside, the men and women of the collective farms, the state-farm workers, all people engaged in agriculture. We must raise agriculture to a level where it would constantly keep pace with industry.

The spring is a decisive time for farm work. In the third year of the Seven-Year Plan period we must do especially well in showing our possibilities of elevating agriculture. All the people engaged in agriculture must apply a maximum of effort to make agriculture meet the growing requirements of the people more fully.

Comrades, there are many wonderful pages in the annals of our country. They are annals written by the labour, the inspiration, the talent, the perseverance and courage of millions of Soviet people.

May our splendid Soviet people, the creators of a new life, the creators of communism, live and flourish! (*Stormy applause.*)

May our socialist country, the country where the Great October Revolution ushered in the new epoch in the development of mankind, live and flourish! (*Stormy applause.*)

Glory to Vladimir Ilyich Lenin, the great leader and founder of the Communist Party and our socialist country! (*Stormy, prolonged applause, cheers.*)

Lenin's genius lights our road to communism, inspires us to perform fresh feats on behalf of peace and the happiness of all mankind! (*Stormy applause.*)

Long live the peoples of the Soviet Union, the builders of communism! (*Stormy, prolonged applause, cheers.*)

DECREE
OF THE PRESIDIUM OF THE SUPREME SOVIET OF THE U.S.S.R. CONFERRING THE TITLE OF HERO OF THE SOVIET UNION UPON THE SOVIET PILOT-COSMONAUT MAJOR Y. A. GAGARIN, THE WORLD'S FIRST SPACE FLYER

For a heroic exploit, the first flight into outer space, that has brought glory to our socialist country, for courage, valour, fearlessness, and selfless service to the Soviet people, to the cause of communism, and to the cause of human progress, the title of HERO OF THE SOVIET UNION is conferred upon the world's first pilot-cosmonaut, Major Yuri Alexeyevich GAGARIN with the award of the Order of LENIN and the GOLD STAR Medal, and a bronze bust of the Hero is to be erected in the city of Moscow.

L. BREZHNEV
Chairman of the Presidium
of the Supreme Soviet of the U.S.S.R.

M. GEORGADZE
Secretary of the Presidium
of the Supreme Soviet of the U.S.S.R.

April 14, 1961
The Kremlin, Moscow

DECREE
OF THE PRESIDIUM OF THE SUPREME SOVIET OF THE U.S.S.R INSTITUTING THE TITLE OF PILOT-COSMONAUT OF THE U.S.S.R

The title of PILOT-COSMONAUT OF THE U. S. S. R. is instituted in honour of man's first flight in space in a satellite spaceship.

L. BREZHNEV
Chairman of the Presidium
of the Supreme Soviet of the U.S.S.R.

M. GEORGADZE
Secretary of the Presidium
of the Supreme Soviet of the U.S.S.R.

April 14, 1961
The Kremlin, Moscow

DECREE
OF THE PRESIDIUM OF THE SUPREME SOVIET OF THE U.S.S.R. CONFERRING THE TITLE OF PILOT-COSMONAUT OF THE U.S.S.R. UPON MAJOR OF THE AIR FORCE Y. A. GAGARIN

For making the world's first space flight in the satellite spaceship *Vostok*, the title of PILOT-COSMONAUT OF THE U. S. S. R. is conferred upon citizen of the Soviet Union, Major of the Air Force Yuri Alexeyevich GAGARIN.

L. BREZHNEV
Chairman of the Presidium
of the Supreme Soviet of the U.S.S.R.

M. GEORGADZE
Secretary of the Presidium
of the Supreme Soviet of the U.S.S.R.

April 14, 1961
The Kremlin. Moscow

TRIUMPH OF LABOUR, SCIENCE AND REASON

PRESS CONFERENCE AT THE HOUSE OF SCIENTISTS

On April 15, the Academy of Sciences of the U.S.S.R. and the Ministry of Foreign Affairs of the U.S.S.R. held a press conference devoted to the successful completion of history's first space flight, undertaken by a Soviet man in the spaceship *Vostok*.

Soviet and foreign correspondents, the diplomatic corps, members of the Presidium of the Academy of Sciences of the U.S.S.R., prominent scientists and representatives of public organisations of Moscow were invited. Altogether, nearly a thousand people came to this memorable press conference.

Yuri Alexeyevich Gagarin, first Pilot-Cosmonaut of the U.S.S.R. and glorious son of the Soviet people, was given a stormy ovation by the correspondents and other participants in the conference.

The conference was opened by Academician A. N. Nesmeyanov, President of the Academy of Sciences of the U.S.S.R.

SPEECH BY A. N. NESMEYANOV

On April 12, 1961, for the first time in history, a spaceship, the *Vostok*, with space flyer Yuri Alexeyevich Gagarin on board, was placed in orbit round the earth in the U.S.S.R.

The event took place in the morning. The spaceship entered an orbit with a perigee of 175 kilometres and an apogee of 302 kilometres above the surface of the earth. The spaceship's orbital period equalled 89.1 minutes. Together with the spaceman it weighed 4,725 kilograms.

The spaceship was fitted with everything necessary to ensure the spaceman's safety during the flight and his safe landing. Many of the systems on board were duplicated. The spaceship carried instruments that allowed the pilot to determine his position on the orbit at any time.

Two-way radio communication was constantly maintained with the cosmonaut during the preparations for the take-off and during the flight itself.

Special tribute must be paid to the courage, endurance, and self-control of the cosmonaut, Yuri Alexeyevich Gagarin. The night before the flight he slept soundly as his doctors ordered him to and was awakened several hours before the take-off. His pulse was 70-75 per minute during the entire period of the preparations for the flight and after the rocket took off. He joked and his cheerfulness gave added confidence that the flight would be successful.

When he was informed that the signal for the start of the rocket's engines was about to be given, he exclaimed cheerily: "Well, here we go!"

During acceleration, while the ship was being placed in orbit, when the powerful rocket engines were working and the spaceman had to endure considerable G-forces, vibration, and noise, Yuri Alexeyevich Gagarin uninterruptedly transmitted all the necessary information regarding his condition and the operation of the ship's cabin systems. After passing through the dense layers of the atmosphere, when the cosmonaut saw the earth, he transmitted: "It's breath-takingly beautiful!"

During the flight Yuri Gagarin was in constant touch with the earth. At 0952 hours, while flying over South America, he reported: "The flight is proceeding normally. I feel well." At 1015 hours, when he was over Africa, he reported: "I feel no ill effects from weightlessness."

The spaceship's deceleration engine was switched on at 1025 hours, and together with its pilot, Major Gagarin, the ship began its descent from the orbit for a landing in a predetermined region. The Soviet ship *Vostok* landed safely at 1055 hours.

A magnificent exploit has thus been performed and an immortal page has been added to the history of human civilisation. This is an exploit of the Soviet people led by our beloved Communist Party and Soviet Government. It is an exploit of large bodies of scientists, designers, engineers, technicians and workers, an exploit of all testers who ensured the faultless preparation and launching of the spaceship, an exploit of all the services that ensured the normal flight of the spaceship and its landing, an exploit of an intrepid son of the Soviet land, Yuri Alexeyevich Gagarin. His name has already become a legend.

This exploit is symbolical in all its aspects: the fact that a Soviet man was the first space pilot, the fact that the first spaceship, in which Yuri Gagarin made his flight, was named *Vostok* (East), and also the fact that the flight was made in the morning. That morning heralded the beginning of a new era.

For all time to come April 12, 1961, will now be linked up with the exploit performed by Yuri Alexeyevich Gagarin. The entire flight round the earth was completed in 108 minutes and these minutes astounded the world.

The culture of mankind has a long and fabulously remarkable history. Each of its exploits, whether it was the invention of the first characters of a written language, the creation of the first steam engines, or the first voyage round the world, was a date when mankind rose to a new stage, reasserting the force of progress and creation. Not all of these exploits were appreciated at once. There was a fierce struggle between the old and the new. And the more revolutionary the events that opened the road to the future were, the more violently were they resisted by the past.

On the threshold of the twentieth century mankind was shown the road to the stars by a man of genius, Tsiolkovsky, whom nobody recognised. His works contain the principles of space science, one of whose brilliant triumphs we are marking today.

The words, "The earth is a cradle of wisdom, but we cannot always live in a cradle," uttered by Konstantin Eduardovich Tsiolkovsky have come true.

Yuri Alexeyevich Gagarin, the first spaceman, has undergone long and strenuous training. It was an unusual, profoundly scientific system of training, which gave the cosmonaut the necessary technical knowledge of the arrangement of his ship and its systems, and a knowledge of astronomy, geophysics, biology, and other sciences.

The pilot underwent acceleration tests in special centrifuges and on vibration stands. Experiments in confined cabins that were the exact replicas of the cabin in a spaceship, lasted for days and weeks. A system of landing was perfected. All this colossal work was crowned by the first space flight in history.

Dear Yuri Alexeyevich, on behalf of the Presidium of the Academy of Sciences of the U.S.S.R. I greet you, a wonderful Soviet man, the Columbus of outer space. (*Prolonged applause. The participants in the conference rise as they greet Yuri Gagarin.*)

Centuries will pass but your name will always remind

men of a great exploit performed by Soviet scientists, designers and you, personally, who have made the first manned flight into outer space. (*Applause.*) You have shown the whole of humanity an example of courage, valour, and heroism in the service of mankind.

Academician A. N. Nesmeyanov presented to Yuri Gagarin the gold Tsiolkovsky Medal, which the Presidium of the Academy of Sciences awarded to the hero for making the world's first flight into space in the ship *Vostok*.

* * *

The floor was given to Yuri Alexeyevich Gagarin. The correspondents rose to their feet and warmly greeted the man whose name has become a legend.

SPEECH BY Y. A. GAGARIN

Dear comrades, esteemed guests,

Many people want to know my biography. Reading the newspapers I find that there are frivolous people in the United States of America, distant relatives of the princes Gagarin, who claim that I am some sort of a relative of theirs. But I'll have to disappoint them. They made a thoughtless and silly claim. I am an ordinary Soviet man. I was born on March 9, 1934, and my father is a collective farmer. My place of birth is Gzhatsk District, Smolensk Region. I do not know and never heard of princes or people of noble birth being among my relatives. Prior to the Revolution my parents were poor peasants. My grandfather was also a poor peasant and there were no princes among us. (*Applause.*) I convey my regret to these noble "relatives", but I shall have to disappoint them. (*Laughter. Applause.*)

I went to school, a trade school in the town of Lyubertsy, Moscow Region, and then to an industrial school in Saratov, where I learned the trade of foundryman. But it has always been my dream to become a pilot and to fly. In 1955 at the same time that I finished the industrial school I completed the courses at the Saratov flying club, after which I was admitted to the Orenburg flying school, which I finished in 1957 and became a military fighter-plane pilot. I served in a unit of the Armed Forces of the Soviet Union.

On my own insistent request I was included among the candidates aspiring to become cosmonauts. I was selected and, as you can see, I have become a space pilot. (*Applause.*) I underwent the corresponding training, the programme for which was worked out by our scientists and which has been described in great detail by the President of the Academy of Sciences. I mastered the instrumentation and was ready for a flight into space.

I am very happy and am boundlessly grateful to our Party and our Government that this flight was entrusted to me. I made it in the name of our country, in the name of the entire heroic Soviet people, in the name of the Communist Party of the Soviet Union and its Leninist Central Committee. (*Applause.*)

I felt very fit and in fine condition before the flight and was absolutely sure that it would be successful. The equipment was very good and very reliable and I and all my comrades—scientists, engineers and technicians—had no doubt that this space flight would be successful.

I felt very well during the flight as well.

In the active sector, during the time the ship was placed in orbit, the acceleration, vibration, and other stresses did not act oppressively on my condition and allowed me efficiently to carry out the programme that was set for the flight.

A state of weightlessness set in after the ship was placed in orbit and the carrier rocket separated. At first this feeling was a little odd although previously I had experienced weightlessness for short periods. But I soon grew accustomed to this state of weightlessness, adapted myself to it and continued carrying out the programme assigned for the flight. It is my own subjective opinion that weightlessness does not affect man's capacity for work or the human organism's ability to carry out its physiological functions.

Throughout the flight I worked fruitfully in accordance with the programme. I ate, drank, and kept up radio communication by telephone and telegraph with the earth through several channels. I watched the work of the equipment on board, reported back to earth and recorded data in the ship's log and on a tape-recorder. I felt well throughout the entire period I was in the state of weightlessness and had full command of my capacity for work. Then, the command for the descent was given at a definite time in conformity with the programme for the flight. The retro-motors were switched on and the ship was given the speed needed to return it to earth. The descent, which was provided for in the programme, was accomplished and I met our beloved Soviet people with delight. The landing was made in a predetermined region.

I would like to say a few words about my observations in space.

The earth can be seen excellently from an altitude of 175-300 kilometres. Its surface has approximately the same appearance as when it is seen from jet aircraft flying at high altitudes. Big mountain ranges, big rivers, large tracts of forest, the shore line and islands can be made out distinctly. The clouds covering the earth's surface and the shadows thrown by these clouds on the earth can also be seen very well. The sky is pitch black. The stars are brighter and more clear-cut against the background of this black sky. The earth has a very characteristic and very beautiful blue halo. The halo is seen very well when you observe the horizon, where the colour of the atmosphere gradually changes from a light-blue, blue, dark-blue and violet into the inky black of the sky. It is a very beautiful sight.

From the shadow I emerged into the sun, which penetrated the earth's atmosphere. Now the halo was somewhat different in colour. A bright-orange, which turned into all the colours of the rainbow: into blue, dark-blue, violet and the black of the sky—could be observed at the very surface, at the very horizon of the earth's surface.

The entry into the shadow of the earth was very rapid. There was immediate darkness and nothing could be seen. During this period I did not observe anything on the earth's surface. Nothing could be seen because, evidently, I was passing over an ocean. Had there been big cities I would probably have seen lights.

The stars can be seen very well. The egress from the earth's shadow is also very rapid and abrupt.

I bore the influence of the space flight very well because I had been fully prepared. At present I am in excellent health.

I am very grateful to our Soviet designers, engineers, and technicians, to the whole Soviet working people, who created the wonderful ship *Vostok*, its marvellous equipment and the splendid powerful carrier rocket that makes it possible to place such huge ships in orbit.

I am boundlessly happy that my beloved country was the first in the history of mankind to penetrate outer space. The first aircraft, the first sputnik, the first spaceship, and the first manned space flight are the stages of the great road that my country has traversed towards the unravelling of Nature's mysteries. Our people have been and are being confidently led towards this goal by the Leninist Communist Party. (*Applause.*)

At every step of my studies, life and work—at the trade school, at the industrial school, at the flying club and at the flying school—I constantly felt the care and attention of the beloved Party, whose member and son I am. I should like to note, in particular, the loving human concern that is shown in the Soviet Union for ordinary folk by the Central Committee of the Party, the Soviet Government and our dear Nikita Sergeyevich Khrushchov. (*Applause.*) Virtually a few minutes after I landed on my beloved Soviet land I received a very warm telegram of congratulations from Nikita Sergeyevich, in which he congratulated me on the successful completion of this space journey. We have dedicated our flight to the heroic Soviet people, to our Government, to our beloved Communist Party and to the Twenty-Second Congress of the Communist Party.

We plan to fly many times, to fly confidently, and really to conquer outer space. (*Applause.*) We are always happy to see achievements in the development of science in other countries and shall be happy to greet the cosmonauts of other countries in outer space. We wish them every success in the peaceful conquest of outer space, and desire to co-operate with them in peacefully utilising outer space. (*Applause.*)

Personally, I would like to make many more flights into space. I liked flying in it. (*Applause.*) I would like to fly to Venus, to Mars, to do some real flying. (*Applause.*)

QUESTION PERIOD

Those present showed great interest in the details of the flight made by the first space pilot. Many questions were put to him and to the Soviet scientists present at the press conference.

Academician Nesmeyanov said that a number of the notes handed up asked whether there had been any preliminary attempts to send a man into space. The writers of the notes referred to reports that had appeared in the Western press.

The President replied, saying that no such attempts had been made. Yuri Gagarin was the first and his attempt had been crowned with success. (*Applause.*)

In answer to the question as to whether it was necessary to send a man into space, Academician Nesmeyanov said: "If there had been no such flight how would man be able to reach distant planets in the future?"

Another question was the following: "Nikita Sergeyevich Khrushchov said that this flight into space was only the first Soviet swallow. When is the next to be expected?" "Swallows usually start their flight in spring," answered Academician Nesmeyanov.

It was asked why the Soviet Union was ahead of the U.S.A. in space research.

"There are many reasons," answered Academician Nesmeyanov, "as there are for any complex phenomenon. Way back in pre-revolutionary Russia, Tsiolkovsky was the first to produce a theory of space travel. The main reason, however, is the possibility of organising and planning scientific and technical work in a socialist state with greater efficiency than in a state with private property and numerous conflicting interests. . . ."

The space pilot himself then went on to answer questions, without constraint, concisely and wittily. He said that the landing technique developed in the Soviet Union had a number of variants, including landing by parachute. In the present case the landing had been effected in the following way: the pilot sat in his cabin and the descent was accomplished successfully, showing the high degree of efficiency of the landing mechanism as a whole.

Question. Will photographs of the earth's surface taken from the *Vostok* be published?

Gagarin. The *Vostok* did not carry a single camera or any other device for photography. No photographs were taken and so there is nothing to publish.

Answering a question on his "way of life" during the flight, Gagarin said, in particular, that he had not experienced hunger or thirst during the flight.

Question. When were you told that you were the first candidate for a space flight?

Gagarin. I was informed in good time that I was the first candidate. (*Laughter, applause.*)

Gagarin spoke of the important part played by radio communication during the space flight. It kept the pilot in constant touch with the earth, enabled him to receive orders, transmit information from the spaceship on the work of the various systems, and gave him a sense of the support of our people, Government and Party so that he did not feel alone during the flight.

The space pilot then told the audience that the landing and the arrival of the ground team at the landing place had been almost simultaneous.

Journalists were interested in Yuri Gagarin's weight. He said that before the flight he had weighed 69.5 kilograms and still weighed the same.

Speaking of the distance travelled in making the descent, Yuri Gagarin said that it was several thousand kilometres,

The retro-motor had been switched on at 1025 hours and the landing had taken place at 1055 hours.

Question. Did you make any preliminary flights on ballistic rockets?

Gagarin. No, I did not.

Question. If you, a family man and the father of two children, were sent into space, the Government and you must have been confident that the flight would have a favourable outcome?

Gagarin. I would like to change the word "sent" for the word "entrusted". I am very glad and proud that I and nobody else was entrusted with this flight. That everything would "work smoothly" and the flight would be a success, nobody doubted—neither the Government, nor the scientists, nor the engineers, nor I. (*Applause.*)

Gagarin was also asked whether he wore a lucky charm or took photographs of his relatives on the flight with him.

Gagarin. I do not believe in omens, lucky charms and that sort of thing. I did not take any photographs with me because I was sure I would return to earth and see my family. (*Applause.*)

Question. Can the spaceship or any parts of it be used for another flight?

Gagarin. That question is more within the competency of our technicians and engineers. But I don't think I am making a mistake in saying that the entire ship and its equipment can be used for another flight into outer space. (*Stormy applause.*)

An inhabitant of South America present at the press conference asked what that continent looked like to Gagarin.

"It was very beautiful," answered Gagarin. (*Applause.*)

In answer to one of the questions, the space pilot said that the Soviet Union is training space pilots in accordance with the space research programme. His words, "I think there will be enough of them to accomplish flights into outer space", were greeted with applause.

Asked whether he had fulfilled the flight programme in full, Gagarin answered in the affirmative.

Question. Would a lengthy stay in outer space be accompanied by any discomfort for the pilot?

Gagarin. The time I spent in orbit allows me to draw the subjective conclusion that a much longer flight could be made.

Question. Does your flight serve to strengthen your political convictions? Does it confirm the idea you expressed that complete and controlled disarmament must be achieved? Why?

Gagarin. I find it difficult to add anything to what Academician Fyodorov has said on that score. I think he answered that question in full.

Answering a question regarding his feelings when he returned to his native earth, Yuri Gagarin said:

"It is hard to express what I feel: the joy, pride, and happiness that a space flight has been accomplished, that I have carried out my assignment successfully. My feeling is one of happiness that the flight has been accomplished by the Soviet Union and that advanced Soviet science has made further progress."

Question. What is your salary and did you receive any special award for the flight?

Gagarin. My salary, like that of all Soviet people, is sufficient to satisfy all my needs. I was awarded the title of Hero of the Soviet Union and that is the highest award in our country.

Question. Do you think you will fly into space a second time or will it be someone else?

Gagarin. I have already stated that I am ready to fulfil any new task allotted by the Communist Party, the Government, and the people. I shall be glad and grateful if I am entrusted with a second flight, but we have many others anxious to make such flight.

The last question put to Yuri Gagarin was: "Could you have flown to the moon on the *Vostok*?

Gagarin. The spaceship *Vostok* was not designed for flights to the moon. Special new ships will be built in our country for that purpose.

The press conference lasted two hours.

MAN'S FIRST FLIGHT INTO SPACE

Man's first space flight in history was made in the Soviet Union on April 12, 1961.

The spaceship *Vostok* with Y. A. Gagarin, Pilot-Cosmonaut of the U.S.S.R., on board was put into orbit round the earth. Without the last stage of the carrier rocket the spaceship weighed 4,725 kg. According to ascertained data obtained by checking up on all measurements, the perigee was 181 km., the apogee 327 km., and the inclination of the orbit 65°4'.

After its flight along the orbit the satellite spaceship safely landed in the predetermined location of our country.

The first space flight made by a Soviet citizen opens up an era of man's direct penetration into space and constitutes one of the major events in the history of civilisation.

The successful flight was a result of an extensive and purposeful programme of space research carried out in the Soviet Union.

The great dream of K. E. Tsiolkovsky, father of space flying, is coming true. Tsiolkovsky said: "Humanity will not stay on the earth for ever, but, in pursuit of light and space, will at first meekly penetrate beyond the atmosphere and will then conquer all of circumsolar space."

DECISIVE STEP IN MASTERING SPACE

For a period of many millenniums man's inquisitive mind strove to penetrate into the depths of the Universe. This expresses man's unquenchable thirst for knowledge, his striving to understand his role in the Universe and to learn to govern the laws of Nature.

Modern science has many means for space research. The distances accessible to these means are expressed in astronomical figures.

Space is a world of stars, stellar associations and galaxies, one of which contains our solar system. Advanced science armed with the theory of dialectical materialism asserts the existence of numerous worlds in which the development of life, the highest form of matter, is possible. The emergence of life in the Universe is in no way an exceptional phenomenon. We cannot say concretely where, in addition to our solar system, there is life today, nor in what form it exists, but it does exist.

With the appearance of man on earth a qualitatively new stage of development of the earth as a planet began. Learning the laws of nature man started changing the earth by arming himself with powerful means in the struggle against nature. From the stone axe man came to perform his greatest feat—the first flight into space.

By flying into space man is directly penetrating into a new sphere, and each penetration into a new sphere involves new discoveries which are frequently unpredictable. Thus, only the flights of the first artificial satellites made it possible to discover the existence of radiation belts around the earth which essentially altered our ideas about the space near the earth and the radiation hazard in space flying.

It is now difficult to estimate the full importance of space flying and the prospects it opens. One thing is certain—

man's penetration into space will immeasurably extend the boundaries of our knowledge and will infinitely enrich our science and culture.

In our days science and engineering are developing at an increasingly rapid pace. Today we witness achievements which could not be conceived some 15-20 years ago. It cannot be doubted that the further development of science and engineering, particularly rocket engineering, will proceed at a constantly increasing rate.

Utilisation of space vehicles for the solution of a number of practical problems can be expected already in the nearest future. The weather and ice reconnaissance services, relay of television and radio broadcasts, and very extensive research outside the earth's atmosphere will be but the first steps along this way. These will be followed by man's flights to the moon and the other planets of the solar system, creation of habitable interplanetary stations and man's gradual adaptation to life in space. In the distant future man will be able—however fantastic it may seem—to establish communication with other worlds.

In preparing and sending a man into space, one of the basic problems with which Soviet scientists and designers were faced was that of ensuring the necessary conditions for his safe flight and return to earth. To solve this problem required a large number of designs and experimental launchings.

On considering the possible variants of man's first flight it was found expedient to carry it out in a spaceship, since such flight directly opens to man the way into space. A flight in a rocket along a ballistic trajectory, which is essentially not a space flight but merely an attempt to create a sensation, was rejected. It is therefore no accident that Soviet scientists and designers directed their efforts, from the very beginning, to developing large and heavy artificial satellites and spaceships. This was the fundamental line of development of space rocketry in the U.S.S.R. Only in this way was it possible to solve the historic problem of man's flight into space.

Beginning with Sputnik II which carried an experimental animal—the dog Laika—on board and up to the spaceship *Vostok* Soviet scientists and designers steadfastly followed this line.

It was necessary to obtain as much information about the work of the spaceships and their equipment as possible, and to develop reliable flying control systems. Development of systems of spaceship orientation and solution of the problem of return to earth were fundamentally new problems.

To make the flight of man on board the spaceship possible it was also necessary to ensure the maintenance of normal pressure, temperature, composition of the air, and other conditions for man's vital activities.

Space research, in addition to the solution of the principal problems in space physics, has furnished the necessary material on the effect of various radiations on the living organism and the meteorite hazard during space flying. Measures to protect the space vehicles against radiation were adopted on the basis of the information obtained.

The extensive experimental material obtained as a result of the flights of the first Soviet satellite spaceships and the development of systems which ensure their safe return enabled the Soviet scientists and designers to start developing a ship for man's flight into space. The extensive and strenuous work resulted in the construction of the spaceship *Vostok*. The last two test launchings of this ship were carried out in March 1961. During these launchings the pilot's seat was occupied by a dummy. In addition, the cabin carried experimental animals—dogs Chernushka and Zvyozdochka.

The flights were made according to the same programme as was planned for the first flight of the ship with a space pilot on board. Both flights, carried out in precise conformity with the pre-set programme, confirmed that the ship and all its systems were very reliably constructed.

The thorough preliminary testing of the spaceship *Vostok* ensured complete success during its very first flight with a spaceman on board on April 12, 1961.

THE SPACESHIP *VOSTOK*

The spaceship *Vostok* was designed on the basis of experience gained in the flights of the first Soviet satellite spaceships.

The satellite spaceship consists of two basic parts:

(1) the cabin of the space pilot, which also houses facilities for human survival and a landing system;

(2) the instrument bay, which houses apparatus that function in orbital flight, and the ship's deceleration motor.

After the ship has been put into orbit it separates from the last stage of the launching rocket. In flight, the on-board apparatus operate according to a definite programme, which covers the measurement of orbital parameters, the transmission to earth of telemetry data and television pictures of the cosmonaut, two-way radio communication with the earth, maintenance of a specified temperature regime on board the ship, and air conditioning in the pilot's cabin. Control of the various components is automatic, by means of on-board programmed systems and, if necessary, by the space pilot.

The programme of the first manned flight was calculated for one circuit of the earth. However, the design and equipment of the satellite spaceship permit longer missions to be carried out.

Upon completion of the flight programme and before re-entry, a special system orients the ship in a definite direction. Then, at a pre-determined point in the orbit the deceleration motor is switched on, cutting the speed of the ship by a pre-computed value. As a result the ship settles into a re-entry trajectory.

The cabin with the spaceman is slowed down in the atmosphere, the flight path being chosen so that re-entry G-forces in the dense layers of the atmosphere should not exceed human tolerance. When the descending vehicle has reached a pre-set height, the landing system is actuated. The actual landing of the pilot's cabin takes place at low speed. The ship covers a total of about 8,000 kilometres between the time the deceleration motor is switched on and impact on earth. The entire period of re-entry lasts approximately 30 minutes.

The outer surface of the pilot's cabin is covered with a layer of heat shield that protects it from burning up during re-entry through the dense layers of the atmosphere. The shell of the cabin has three portholes and two fast-opening hatches. The portholes are fitted with heat-resistant glass and permit the astronaut to make observations during the entire flight.

In the satellite spaceship the pilot occupies an ejection seat, which serves as his work place during the flight and also as a means of escape from the vehicle in case of emergency. The seat is positioned so that the accelerations of powered flight and re-entry act on the cosmonaut in the most favourable direction (from chest to back).

In the first flight, the space pilot was dressed in a protective spacesuit that would save his life and enable him to continue working even in case of depressurisation of the pilot's cabin while in flight.

The satellite spaceship also has the following apparatus and systems:

apparatus and equipment necessary for normal functioning of the human organism (an air-conditioning system, a system of pressure control, food and water, and a system of removal of human waste products);

flight control instruments and a system of manual control of the ship (pilot's panel, instrument board, manual-control unit, etc.);

landing systems;

radio for communication of the pilot with earth;

a system of self-registration of instrument readings, radiotelemetering system, and various pick-ups;

a television system for observing the cosmonaut from the earth;

apparatus registering human physiological functions;

the deceleration motor system of the spaceship;

the orientation system instruments;

flight control instruments;

radio systems for measuring orbital elements;

a thermal-control system;

power sources.

On the skin of the ship are situated: control devices, orientation cells, the louvers of the thermal-control system, and the radio antennas.

The pilot's cabin of the satellite spaceship is much more spacious than that of an aircraft. The equipment of the cabin is designed for the convenience of the cosmonaut. Seated, the space pilot can do all that is required of him: make observations, establish communication with the earth, do flight check-ups, and, in the case of necessity, control the ship.

Built into the pilot's seat are:

a detachable back with a strap system for positioning the body of the pilot for ejection and descent by parachute;

parachute systems;

ejection system;

portable emergency kit (containing food, water and accessories) and radio equipment for communication and direction finding, which the cosmonaut can use after landing;

the spacesuit ventilation system and the parachute oxygen-unit;

automatic system of the seat.

The cosmonaut can land in the cabin of the ship; this landing technique was tested on the fourth and fifth Soviet satellite spaceships that carried experimental animals. Another version of landing that was provided for was the ejection of the seat and the spaceman from the cabin at about 7 kilometres altitude, with subsequent landing with parachutes. This version was likewise tested in launching satellite spaceships.

The air-conditioning system of the spaceship maintains in the pilot's cabin normal pressure, normal concentration of oxygen, with the concentration of carbon dioxide not exceeding one per cent, the temperature at 15-22° C, and relative humidity within 30 to 70 per cent. The air composition was regenerated (carbon dioxide and water vapour absorbed and a corresponding quantity of oxygen released) by means of highly active chemical compounds. The process of regeneration was regulated automatically. When the amount of oxygen decreases and the concentration of carbon dioxide increases, a special sensing device releases a signal that actuates a mechanism which alters the mode of operation of the regenerator. If excess oxygen is liberated, the same mechanism automatically reduces the release of oxygen into the atmosphere of the cabin. The humidity of the air is controlled in like manner.

Special filters are used to purify the air in the case of contamination by harmful impurities released as human waste products and through operation of apparatus.

A thermal-control system is used to maintain a pre-set temperature regime in the ship in flight. A distinguishing feature of this system is the use of a liquid coolant to remove heat from the pilot's cabin. The coolant is maintained at a stable temperature. The coolant enters a liquid-air radiator from the thermal-control system. The amount of air consumed by the radiator is automatically controlled depending on the temperature in the descending vehicle. In this way, the pre-set temperature regime in the cabin is maintained to a high degree of accuracy.

To maintain the coolant at a stable temperature and to ensure the requisite temperature regime in the instrument bay, the outer surface of the latter is equipped with a radiation heat-exchanger having an automatically controlled system of louvers.

Prior to turning on the deceleration motor for re-entry and landing in a predetermined area, the spaceship must be very specifically oriented in space. This is done by means of a system of orientation. In the given flight, one of the ship's axes was oriented in the direction of the sun. This system includes a series of optical and gyroscopic sensing devices. The signals delivered by these devices are converted, in an electronic unit, into commands that control the system of controls. The system of orientation automatically seeks the sun, turns the ship accordingly, and maintains it in the required position to a high degree of accuracy.

After the ship has been oriented, the deceleration motor is switched on at a pre-set time. The commands for switching on the orientation system, the deceleration motor, and other systems are issued by a programmed electronic computer.

The satellite spaceship is equipped with radio-measuring and radio-telemetering systems for measuring the orbital elements of the ship and for controlling the operation of on-board apparatus. Ground stations situated within the U.S.S.R. measure the parameters of motion of the ship and receive telemetry information during the flight. Measurement data are automatically transmitted via communication links to computing centres where they are processed on electronic computers. As a result, during the flight, information concerning the basic orbital elements is received promptly and predictions are made of the flight path of the ship.

The spaceship also has a radio system "Signal" operating on a frequency of 19.995 megacycles per second (Mc/s). This system serves for direction finding of the ship and for transmission of a part of the telemetry information.

The television system transmits pictures of the cosmonaut to earth thus making it possible to observe his condition. One of the television cameras gives a full-face picture of the pilot, while the other gives one in profile.

Two-way communication between the spaceman and the earth is handled by a radio-telephone system operating on short (9.019 and 20.006 Mc/s) and ultrashort bands (143.625 Mc/s) waves.

The ultrashort-wave channel is used for communication with ground stations at distances up to 1,500-2,000 kilometres. Experience shows that over most of the orbit it is possible to maintain communication with ground stations within the Soviet Union over the short-wave channel.

The radio-telephone system includes a tape-recorder that permits recording what the cosmonaut says in flight; it is then played back and transmitted when the ship flies over the ground receiving stations. Radio-telegraph transmission by the cosmonaut is likewise provided for.

The instrument board and the pilot's panel mounted in the cabin are designed to monitor the operation of the basic on-board systems and to ensure, in case of necessity, a manually controlled re-entry of the ship. The instrument board has a number of pointer indicators and signal panels, an electric clock, and also a globe, the rotation of which is synchronised with the orbital motion of the ship. The globe enables the cosmonaut to determine the position of the ship at all times. The pilot's panel contains knobs and switches which control the radio-telephone system, the temperature in the cabin, and turn on manual control and the deceleration motor.

When designing the spaceship, particular attention was paid to ensuring safety in flight. The launchings of the first Soviet satellite spaceships confirmed the high degree of reliability of their components and equipment. However, the spaceship *Vostok* was provided with a range of additional devices to preclude the possibility of any accidents and guarantee safe manned flight. This trend of development is in full accord with the basic task—to create vehicles that enable man confidently to penetrate into outer space.

For orientation of the ship in the case of manual control, the spaceman makes use of an optical orientor that permits determining the position of the ship relative to the earth. The optical orientor is mounted on one of the portholes of the pilot's cabin. It consists of two annular mirror reflectors, a light filter and a grid-covered glass. Rays of light coming from the line of the horizon fall on the first reflector, pass through the glass and thence on to the second reflector, which sends them through the grid-glass into the eye of the cosmonaut. If the ship is properly oriented relative to the vertical, the space pilot sees the horizon, in his field of view, as a ring.

The cosmonaut sees the portion of the earth's surface directly under him through the central part of the porthole. The position of the longitudinal axis of the ship relative to the direction of flight is determined by observing the motion of the earth's surface in the field of view of the orientor.

By manipulating the controls, the cosmonaut can turn the ship so that the line of the horizon is seen in the orientor in the form of a concentric ring, while the direction of motion of the earth's surface coincides with the course line of the grid. This will indicate that the ship is properly oriented. When necessary, the field of view of the orientor may be protected with a light filter or blind.

The globe mounted in the instrument board enables the pilot not only to determine the position of the ship but to predetermine also the point of re-entry when the deceleration motor is turned on at a specific instant of time.

Finally, the ship is designed to ensure re-entry and landing (if the deceleration motor fails) by utilising aerodynamic braking in the atmosphere.

The supplies of food, water, regeneration materials, and the electric power sources are sufficient for flight missions of up to 10 days' duration.

The ship is designed to prevent excess temperature rises in the cabin in the case of continuous heating of the outer surface when the ship is gradually slowed down in the atmosphere.

MEDICAL AND BIOLOGICAL PROBLEMS OF MAN'S FLIGHT INTO SPACE

To solve the problem of man's possible flight into space and of his medical protection it was necessary:

1. to study the effect of the factors of space flying on the organism and to investigate the possible forms and methods of protection against the unfavourable action of these factors;

2. to elaborate the most effective methods of ensuring normal conditions for man's vital activities in the cabin of the spaceship;

3. to elaborate methods of medical selection and training of spaceship crews, as well as a system of continuous medical control of the pilots' health and capacity for work all through the flight.

All of the afore-mentioned problems included a large number of specific objectives whose study and solution engaged, for a period of ten years, the untiring efforts of specialists in physiology, hygiene, psychology, biology, clinical and occupational medicine. The studies were conducted in ground laboratories and during rocket flights of animals. Good use was made of the extensive experience accumulated in the applied fields of physiology and medicine, especially in aviation medicine, and in the protection of health in underwater navigation. Wherever it was thought advisable, special ground stands were developed to study, in laboratory conditions, the effect produced on the organism by factors which come into play during space flights. The effect of accelerations and their tolerance by the organism were studied with the aid of centrifuges. These machines produced the same accelerations that arise during the launching of spaceships and their return to earth.

The effect of other factors on the organism was studied with the aid of vibration platforms, heat and vacuum chambers and other installations. However, laboratory experiments could, as a rule, furnish answers only as regards the effect of some one of the afore-mentioned factors on the organism, while during actual flying in rockets they act in combination and simultaneously. Besides, in laboratories it was impossible to study the behaviour of animal organisms under conditions of weightlessness. The biological studies conducted in rockets since 1951 therefore served as an essential approximation to the study of the effects produced by space flying on the organism.

Dozens of rockets carrying animals were launched on experimental flights to an altitude of up to 450 km. These experiments furnished extensive scientific material showing the reactions of the physiological systems and the behaviour of animals (dogs, rabbits, rats, and mice) during the different stages of the flight. Careful studies of the experimental animals during flying, as well as for a long time after their return to earth, warranted the conclusion that the conditions of rocket flying in the upper layers of the atmosphere are tolerated by animal organisms quite satisfactorily. The changes noted in the different physiological functions during the flight were not of a pathological character; not infrequently they disappeared in the process of the experiment and never reappeared.

However, in view of the fact that the rockets were flown for a short time, it was impossible to investigate the biological effect of such important factors of space flying as long-continued weightlessness and cosmic radiation. The opportunity to utilise artificial satellites of the earth for biological experiments, which presented itself in 1957, constituted an extremely important step forward.

The first of these experiments was carried out on Sputnik II. It not only confirmed and extended the information furnished by the preceding biological experiments in rockets. It demonstrated for the first time that the long-continued state of weightlessness does not in itself affect the vital functions.

The biological experiments were continued on the first Soviet spaceships. The programme of these medical and biological studies included the solution of a number of new problems.

It was important, besides additional and deeper studies of the effect produced on the organism by long-continued weightlessness and the transitional states from weightlessness to accelerations and vice versa, more thoroughly to investigate the biological effects of cosmic radiation. Studies of the peculiarities of the work and efficiency of the systems which would, during future flights, ensure normal conditions for man's vital activities and would guarantee his safe return to earth also constituted an important part of the programme. To carry out the planned programme, various representatives of the organic world on earth, from the simplest forms of life to the higher vertebrates, were placed in the first Soviet spaceships.

Experiments on various species of animals and plants made it possible to study particularly fully and in detail the effect of the conditions of space flying on the most diverse processes and functions of the organism. Very extensive information was gained on the behaviour and physiological functions of the experimental dogs during flying. The behaviour of the animals was observed by means of a special TV system. The analysis of the information obtained has shown that the animals not only fully retain their vital functions under conditions of long-continued weightlessness and the subsequent effect of accelerations, but also that their basic physiological functions are not disturbed. Nor has a sufficiently long and careful study of the animals after their flight revealed any deviations from normal.

Very serious attention was devoted to discovering the possible effects of cosmic radiation during flying in a spaceship. The numerous methods used in the solution of this problem have disclosed no changes which may be attributed to ionising radiation.

The results of the medical and biological studies conducted in spaceships warranted a very important and consequential conclusion. It was recognised that flying in spaceships along orbits below the radiation belts near the earth is safe for the highly organised representatives of the animal world. The results of the biological experiments were utilised in solving the problem of man's tolerance of the flying conditions.

On the basis of this, as well as by taking into consideration the results of the laboratory studies, it was concluded that man may fly in space without jeopardising his health.

TRAINING SPACE PILOTS

The first space flight could be made only by a person who was aware of the enormous responsibility of his assignment and who consciously and voluntarily agreed to exert all his efforts and knowledge and perhaps give his life to perform this outstanding exploit. Thousands of Soviet patriots, people of various ages and occupations, volunteered to fly into space. Soviet scientists were faced with the problem of making a scientific choice of the first spacemen from the large number of volunteers.

During a space flight man would be subjected to the action of many factors of the external environment (acceleration, weightlessness, etc.) and considerable nervous and emotional strain which require man to mobilise all his moral and physical faculties. At the same time the space pilot must retain his capacity for work and the ability to react to any situation during the flight; if necessary he must also take over the controls of the spaceship. All this determined the high requirements made on the spaceman's health, his mental faculties, and the level of his general and technical training.

These qualities are usually found in flyers. The occupation of flyers requires nervous and emotional stability and will power, which are particularly important for the first space flights. Later the category of persons taking part in such flights undoubtedly must and may be considerably extended.

During the organisation of the group of cosmonauts many of those who had volunteered to fly into space were interviewed. Those of the flyers who were best prepared for space flying were given thorough clinical and psychological examination. The aim of this examination was to determine the state of health, disclose the latent insufficiency or lowered resistance of the organism to various factors of a space flight, and to evaluate the reactions to these factors.

The examination was conducted by a number of modern biochemical, physiological, electrophysiological and psychological methods and special functional tests which make it possible to evaluate the reserve possibilities of the organism's basic physiological systems (examination in a pressure chamber under considerable air rarefaction, during changes in barometric pressure and breathing oxygen under increased pressure, examination in a centrifuge, etc.).

Psychological tests aimed at finding persons with the best memory, resourcefulness, active and readily shifting attention, and ability to perform precise and co-ordinated ...ovements formed an important part of the examination.

The clinico-physiological examination resulted in the selection of a group which began to be instructed and trained on special stands and training arrangements which imitate on the ground and in the air the factors of space flying. At the same time the individual peculiarities in the reactions of the organism to the effect of the imitated factors were determined.

The programmes of special instruction were aimed at giving the spacemen the necessary knowledge in the basic theoretical problems connected with the objectives of the forthcoming flight and the practical skills in using the equipment and apparatus installed in the cabin of the spaceship. The programmes included instruction in the following disciplines: fundamentals of rocket and space engineering, construction of the spaceship, special questions of astronomy and geophysics, and fundamentals of space medicine.

The complex of special training and tests included:

flying in planes under conditions of weightlessness;

training in a model of a spaceship cabin and on a special training device;

protracted staying in a specially equipped sound-insulated chamber;

training on a centrifuge;

parachute jumps from planes.

Certain questions of ensuring man's flight into space, particularly questions connected with the space pilot's nutrition and clothing, as well as the system of air regeneration, were also solved in the course of the special training.

The individual reaction of the spacemen to the effect of weightlessness and the transition from weightlessness to acceleration were studied during the flights in planes. The possibility of maintaining radio communication, taking food and water, etc., was also investigated. This made it possible to answer a number of important questions concerning man's possible actions during space flying.

It was found that all the spacemen selected could easily tolerate the state of weightlessness. Besides, it was demonstrated that under conditions of weightlessness lasting up to 40 seconds it was possible to consume liquid, semiliquid and solid food, to perform fine co-ordinated acts (writing and purposeful movements with the hand), maintain radio communication, read and orient oneself in space visually.

The training in the model of a spaceship cabin and on the special training stand was aimed at mastering the equipment and apparatus installed in the cabin, practising the different variants of the flying assignment and habituating the flyers to the real spaceship cabin. A special training stand was developed for this purpose. With the aid of electronic devices this stand made it possible to duplicate on the instruments the changes which actually occur in flight. The pilot also acted as he would in actual flight. The training offered opportunities of imitating unusual (emergency) variants of the flight and taught the space pilot what to do under such circumstances.

The main objective of the examination during the spaceman's long stay in the specially equipped sound-insulated chamber was to determine his neuropsychic resistance during a long stay in a limited isolated space alone with the external stimuli considerably diminished. At the same time his day and diet were planned as though he were actually flying.

A wide range of physiological studies, as well as special psychophysiological methods, made it possible to disclose persons who showed the best indices as to precision and accuracy in fulfilling their assignments, and possessing a more stable neuroemotional sphere.

During the tests (training) on the centrifuge and in the heat chamber the cosmonauts' individual tolerance of the particular influences exerted on them was determined. the effect of these influences on the basic physiological functions was studied and questions of increasing the resistance of the organism to the created factors of the external environment were settled. The studies revealed that the spacemen were capable of resisting the effects of the afore-mentioned factors, and disclosed persons who endured the tests best.

In the course of parachute training each space pilot performed a few dozen jumps. The physical training of the group of space pilots consisted of systematic work-outs and morning setting-up exercises. The systematic work-outs were conducted with due allowance for each spaceman's individual physical development. The morning exercises were performed for one hour every day and were aimed at general physical conditioning. The physical training was designed to enhance the organism's resistance to the effect of accelerations, develop and improve the ability freely to control the body in space, and increase the capacity for enduring prolonged physical strain.

The physical training was conducted under constant medical observation and consisted of specially chosen gymnastic exercises, games, diving, swimming and exercises on special apparatus.

As soon as the programme of special training was carried out, direct preparation for the forthcoming flight into space was organised.

This preparation included:

studying the flight assignments, maps of the landing area, navigation instructions, maintenance of radio communications, etc.;

studying the emergency kit and its utilisation on location after landing, studying the direction finding system, etc.;

tests on the centrifuge in the pressure suit under maximum expected accelerations;

protracted tests in a model of the spaceship with utilisation of all systems ensuring vital activity.

As a result of the instructions and training a group of cosmonauts prepared for flying into space was selected.

Major of the Air Force Y. A. Gagarin was chosen to make the world's first space flight.

Y. A. Gagarin, wonderful Soviet man, was born March 9, 1934, into the family of a collective farmer. He always wanted to become a flyer. After graduation from the Orenburg flying school as a fighter pilot in 1957 he served in one of the units of the Soviet Armed Forces. On his insistent request he was included in the group of space-flying candidates and successfully passed the selection tests. During the training of the group of space pilots Gagarin was one of the best.

Yuri Alexeyevich Gagarin has fully justified the great honour of being the world's first spaceman.

FIRST SPACE FLIGHT

The spaceship *Vostok* took off on April 12, 1961, at *0907 hours* Moscow Time.

During acceleration, space pilot Yuri Gagarin kept in constant radio-telephone contact with the ground flight control station. The cosmonaut felt well throughout this stage of the flight, recording the changes in the G-forces and the moments the stages of the carrier rocket separated from the ship. The noise in the ship's cabin was not greater than the noise in the cabin of a jet aircraft. While still in the acceleration stage Gagarin saw the earth through the portholes.

The ship's apparatus during the orbital flight, its orientation and descent were controlled automatically. However, in case of need, the cosmonaut could at his own discretion or at a command from the earth take over the control of the ship, determine his position and descend to a selected area.

A state of weightlessness set in when the ship entered its orbit. The cosmonaut at first felt odd but soon he adapted himself to it. Throughout the period of weightlessness Yuri Gagarin felt well and had full command of his capacity for work.

In conformity with the task and programme of the flight, he watched the work of the ship's equipment, was in constant contact with the earth by telephone and telegraph, observed his surroundings through the portholes and optical orientor, transmitted his observations to earth and recorded them in the ship's log and on a tape-recorder, ate and drank water.

The earth's surface is clearly visible from an altitude of up to 300 kilometres. Coastlines, big rivers, the relief, tracts of forests, clouds and cloud shadows are seen distinctly. While flying over the territory of the Soviet Union, Yuri Gagarin saw the outlines of collective-farm fields.

The sky is pitch black. Against it the stars are brighter and can be seen more distinctly than from earth. The earth has a beautiful blue halo around it, and the colours on the horizon change from light blue through blue, dark blue and violet to the black of the sky. As the ship emerges from the earth's shadow, a bright orange colour that gradually changes into all the colours of the rainbow is observed on the earth's horizon.

At *0951 hours* the ship's orientation system was actuated, and after the ship emerged from the earth's shadow it orientated the ship's position by the sun,

At *0952 hours* cosmonaut Yuri Gagarin flew over the region of Cape Horn and reported that he was feeling well and that the apparatus on board were operating normally.

At *1015 hours* the automatic programmer issued a signal to have the ship's apparatus ready to switch on the deceleration engine. At this time the ship was approaching Africa and a report was received from Yuri Gagarin on the progress of the flight.

At *1025 hours* the deceleration engine was switched on and the ship entered the descent trajectory.

At *1035 hours* the ship entered the dense layers of the atmosphere.

Completing the world's first space flight with a cosmonaut on board, the spaceship *Vostok* landed in a predetermined area at *1055 hours* Moscow Time.

Pilot-cosmonaut Yuri Gagarin felt well after his return from his flight in space. His health showed no aftereffects.

History's first flight in outer space, accomplished by the Soviet cosmonaut Yuri Gagarin in the spaceship Vostok, *has made it possible to draw the immensely important scientific conclusion that manned flights in space are practicable. It demonstrated that man can normally bear up against the conditions of a space flight, the placing of a ship in orbit and the return to earth. This flight showed that in a state of weightlessness man fully retains his capacity for work, his co-ordination of movements, and his clarity of thought.*

Extraordinarily valuable information was obtained on the effectiveness of the spaceship's design and equipment in flight. It was fully confirmed that the scientific and technical solutions adopted for the spaceship's design were correct. The reliability of the carrier rocket and the excellent design of the spaceship were also confirmed.

We now have the means for manned flights in space,

The first manned flight in space opens a new, space era in the history of mankind.

The time has now come to implement projects that formerly seemed fantastic—the creation of extraterrestrial scientific observatories and journeys by man to the moon and to Mars, Venus, and other planets of the solar system.

The new space era in the history of mankind will witness a colossal extension of man's sphere of life and activity and the conquest of circumsolar space.

ERRATA

Page	Line		
18	7	*For* "middle peasants" *read* "lower middle peasants"
21	5	*For* "two hundred households" *read* "one hundred households"
24	10-11	*For* "After another few years," *read* "Then, after a number of years,"
38	9-12	*For* "who are able to work but remained out of job before the organization of people's communes," *read* "who are able to work but had not yet taken up productive work before the organization of people's communes,"
53	14-15	*For* "appropriately rearrange the taxes due the communes." *read* "appropriately readjust the taxes due from the communes."
53	15	*For* "transition" *read* "gradual transition"
58	16	*For* "rich peasants" *read* "well-to-do peasants"
58	21-22	*For* "Socialist activists" *read* "Socialist-minded and proficient activists"
59	4	*For* "and checking up work done." *read* "and business accounting."
65	1 (footnote)	*For* "mineral oil products" *read* "oil-bearing crops"
89	17	*For* "in the co-operative" *read* "during the merger of the co-operatives"

Milton Keynes UK
Ingram Content Group UK Ltd.
UKHW011431220424
441559UK00001B/90